This book is printed on acid-free paper.

Copyright © 2025 Zouev IB Publishing. All rights reserved.

No part of this book may be used or reproduced in any manner whatsoever without written permission, except in the case of brief quotations embodied in critical articles or reviews.

Published 2025

Printed by Zouev IB Publishing

ISBN 978-1-0684442-3-4, paperback.

IB MATHEMATICS IA

THE COMPLETE GUIDE TO THE INTERNAL ASSESMENT

Robert Flynn (Mr. Flynn IB)

IB Mathematics Video Lessons.
IA Guidance. Past Paper Solutions.

Table of Contents:

Introduction 1

Section 1 .... **3**

Chapter 1: Understanding the IA 5

Chapter 2: Choosing a Topic 9

Chapter 3: The Aim 25

Chapter 4: Do the Math 29

Chapter 5: The Criteria 33

Chapter 6: Tips from the Principal Examiner 41

Chapter 7: The Writing Process 47

Chapter 8: Final Checklist 51

Section 2 .... **55**

Example 1: Table Tennis Shot 59

Example 2: Dubai Ferris Wheel 77

Example 3: Volume of an Egg 91

Example 4: Turtle House Area 109

Example 5: Cake Optimisation 127

Example 6: Swimming Normal Distribution 143

Example 7: Russian Dolls 159

Example 8: Guitar Area 183

Example 9: Triathlon Progress 213

Example 10: Angry Birds 233

Example 11: Voronoi and Travelling Salesman Problem 247

Example 12: Gini Coefficient 271

Example 13: Diwali Lights 287

Introduction

Welcome to your complete guide to the IB Mathematics Internal Assessment (IA).

If you feel overwhelmed, stuck for ideas, or unsure where to even start, you are not alone. The IA can feel like a huge task. But it doesn't have to be. This guide is designed to walk you through the process step-by-step, with practical advice that's clear, to the point, and built from real examiner experience.

I'm Mr. Flynn IB. I'm an IB Maths Examiner and Moderator, and I have seen over 1000 IAs firsthand. Through my YouTube channel and my website I've helped students all over the world navigate the IA process. I know where students lose marks, where they get stuck, and how they can avoid the most common pitfalls. I also know what a good IA looks like — and how to get there.

The best part? You don't have to be amazing at maths to do well in the IA. It's worth **20%** of your final grade, and it rewards thoughtfulness, creativity, and effort — not just exam technique. I once had a student who scored a 3 in their final exams, but got 19 out of 20 in their IA, and ended up with a 5 overall. The IA is a real opportunity especially if exams aren't your strength.

This book is divided into two main sections:

- **Section 1** is your IA Guide. It will take you from understanding what the IA is, all the way through to submitting a finished piece of work that has the potential to score a 7. You'll learn how to choose a topic, write a clear aim, apply the right mathematics, and meet the assessment criteria in a way that examiners reward.
- **Section 2** gives you **13 sample IAs** with commentary. These real examples will show you what works, what doesn't, and how examiners think when they read your work.

These sections serve different purposes. Section 1 is designed to be read in order. It's your step-by-step guide. Section 2 is more like a library. You don't need to read every sample from start to finish. Instead, dip in and out. Use them for inspiration, to check how certain topics were approached, or to better understand what strong IAs look like in practice.

By the end of the guide, you will not only know how to write a strong IA, you'll understand exactly what it takes to achieve a 7.

Let's get started.

Section 1

The IA Guide

Chapter 1: Understanding the IA

Before you can write a high-scoring IA, you need to understand what the IA actually is. This isn't a typical test or a project where you simply apply a formula and get a right answer. The Maths IA is your opportunity to explore a mathematical idea that interests you, investigate it deeply, and communicate your process clearly.

In this chapter, we'll break down the basics of the IA: what you need to produce, how it's structured, how it's assessed, and most importantly, what the examiners are looking for when they read your work.

What is it?

- A short (12-20 pages) written report
- An Individual piece of work
- You will investigate an area of mathematics that interests you showing your process, your understanding and your reflections
- You choose the topic
- It is worth 20% of your overall grade
- A good IA will take a minimum of 20 hours
- You get written feedback on 1 draft
- Marked by your Teacher
- Moderated by the IB

What's the Goal?

The IA isn't about solving a problem with a correct answer. It's about:

- Choosing a mathematically rich idea that interests you,
- Investigating it in a logical and creative way,
- Showing clear mathematical thinking, and
- Reflecting on your process and findings.

If you do this well, even with a relatively simple topic, you can score very highly.

Structure

Your IA should be a between **12 and 20 pages**, and that includes:

- Text
- Graphs
- Equations
- Tables
- Diagrams

Keep it concise. Every sentence should either explain something or move your investigation forward. No filler!

There's no fixed format, but most strong IAs follow a structure something like this:

1. **Introduction**
2. **Mathematical Process**
3. **Interpretation**
4. **Conclusion**
5. **References**
6. **Appendix**

Don't worry, we'll go into detail on each of these sections in later chapters.

What Are Examiners Looking For?

Your IA will be marked out of 20, split into five criteria:

Criterion	Marks	Description
A Presentation	4	Organisation, clarity, structure, flow, coherence
B Mathematical Communication	4	Correct notation, maths steps shown, definitions, correct diagrams and graphs and degree of accuracy
C Personal Engagement	3	Making it your own, thinking independently, originality, curiosity, engagement with the maths

Criterion	Marks	Description
D Reflection	3	Review, analysis, evaluation, limitations and assumptions, critical thinking, strengths and weaknesses
E Use of Mathematics	6	Understanding the maths, relevant to aim, appropriate level

We'll unpack each of these in Chapter 5 but it's important to keep them in mind from the very start.

What's next?

Now that you understand what the IA is and what the examiners are looking for, it's time to focus on the most important decision you'll make: **choosing your topic**.

This can make or break your IA. A strong topic makes the whole process easier and more enjoyable. It sets you up to score highly across all five criteria.

In the next chapter, we'll explore how to find a topic that's interesting, mathematical, and that actually works within the IA framework.

Let's get into it.

Chapter 2: Choosing a Topic

Choosing your IA topic is the **most important decision** you'll make. A strong topic sets the stage for everything else — from your aim to your mathematics to your final score. A poor topic, on the other hand, can turn the entire process into a frustrating slog, no matter how good you are at maths.

I've seen students write a **Grade 7** IA in a weekend once they found the right topic. I've also seen students spend months writing disappointing IAs simply because their topic was vague, unworkable, or too complicated. So, let's get this part right.

In this chapter, you'll learn how to choose a topic that:

- Interests you personally
- Has clear mathematical depth
- Is actually doable within 12–20 pages
- Gives you something real to explore and reflect on

Let's find a topic that makes your IA easier not harder.

What Makes a Good IA Topic?

Here are the two key ingredients of a successful IA topic:

1. **You're genuinely interested in it.**
   You'll be spending hours researching, calculating, writing, and reflecting. If you're bored, it will show.
2. **It has mathematical depth.**
   There needs to be enough mathematics to explore. That means going beyond prior learning or just substituting numbers into formulae. The maths should be central, not just background.

Where to Look for Ideas

Start with your own interests. Some of the best IAs come from simple curiosity:

- A concept you saw on YouTube or TikTok and want to explore further
- A sport or game you love

- A pattern you noticed in everyday life
- Something strange or surprising from your maths class
- A personal hobby like music, fashion, coding, investing, or design

Pro Tip: Watch my YouTube playlist called IB Math IA Ideas.

A good IA doesn't need to be totally original but it does need to be personal. It should feel like *your* investigation, not a recycled school project.

IA Ideas

This section is designed to help spark your thinking and guide you toward a topic that fits both your interests and the expectations of the IB.

First, I'll give you a **list of mathematical topics and concepts** that tend to work well in IAs. Then, you'll see a brief snapshot from a real IA that used each type of maths effectively. Finally, I'll share a longer list of topic ideas — collected over the years — to give you even more inspiration.

These are some of the most commonly used mathematical techniques in successful IAs:

- **Linear Modelling**
- **Quadratic modelling**
- **Polynomial modelling**
- **Exponential modelling**
- **Sinusoidal modelling**
- **Logistic modelling**
- **Ellipse modelling**
- **Voronoi diagrams**
- **Logarithmic graphs (log-log or semi-log)**
- **Area under the curve**
- **Volume of revolution**
- **Optimisation**
- **Normal distribution**
- **Hypothesis testing**
- **Graph Theory**

Remember, it's not about using advanced maths, it's about using maths appropriately and meaningfully. Even simple models can earn top marks if they're used well.

Linear modelling

In this example, a student modelled her triathlon progress using a linear model. See Sample 9.

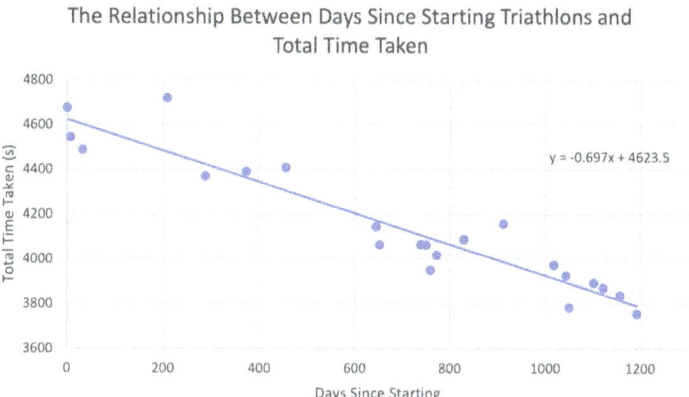

Quadratic modelling

In this example, a student modelled both his and Ray Allen's basketball shot in order to compare.

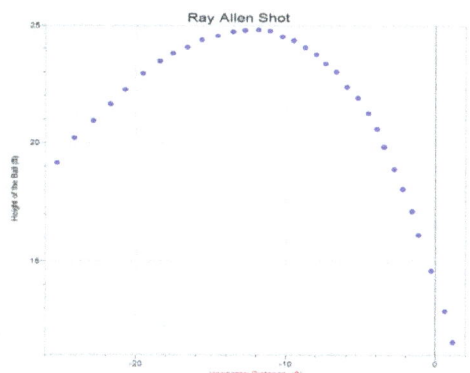

Polynomial modelling

In this example, a student used cubic and quartic polynomial functions to model a guitar from a photo. See sample number 8.

Figure 2.1.4
Annotated Image Showing Fit of Cubic Functions

Exponential modelling

In this example, a student created an exponential model of his coffee cooling to find the optimum time to wait before drinking his coffee.

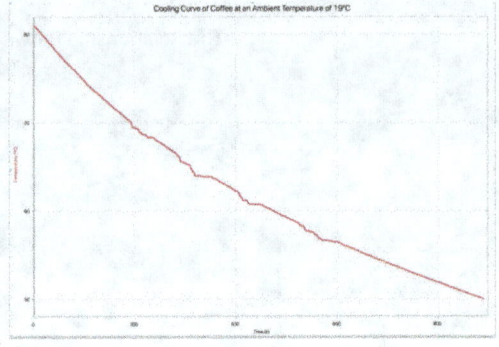

Graph 2: Cooling Curve of Coffee at an Ambient Temperature of 19ºC

Sinusoidal modelling

In this example, a student modelled the height of a pod in the Dubai Ferris Wheel using a sinusoidal function. See sample number 2.

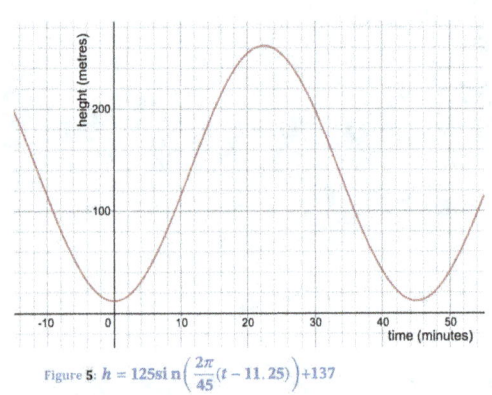

Figure 5: $h = 125\sin\left(\frac{2\pi}{45}(t-11.25)\right)+137$

Logistic modelling

In this example, the student used logistic functions to model the population of China and India to determine when India would become the world's most populous country.

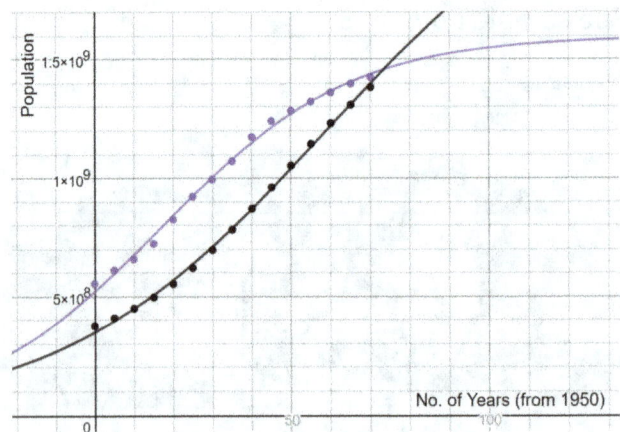

Ellipse modelling

In this example, a student used an equation of an ellipse to model an egg in order to find its volume. See sample number 3.

Image 3: Ellipse equation plotted onto image of Conundrum Egg on GeoGebra

Normal distribution

In this example, a student used normal distribution to find the probability that the boys' basketball team would score a particular number of points.

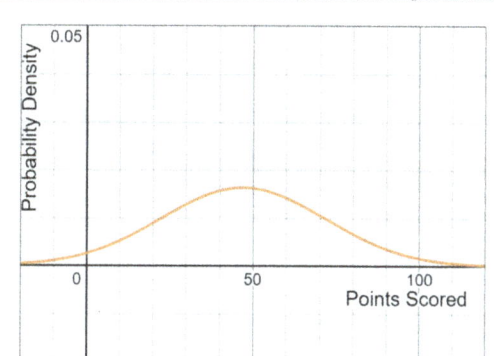

Normal distribution for the boys' basketball team's points scored per match

Area under the curve

In this example a student found the area under pharmacokinetic curves to determine which painkiller provides the most relief.

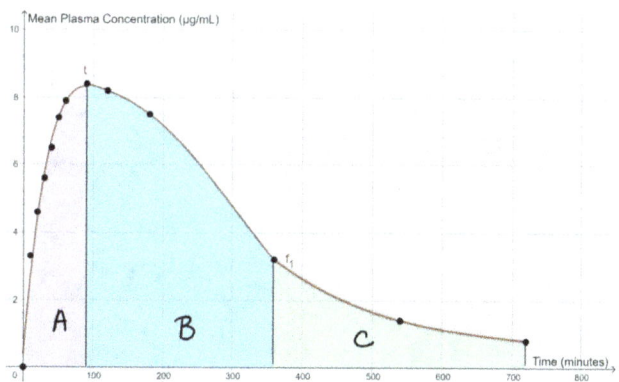

Volume of a revolution

In this example, a student used volume of revolution to calculate the volume of their trombone.

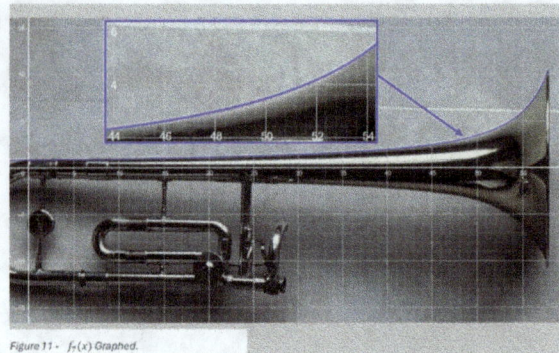
Figure 11 - $f_2(x)$ Graphed.

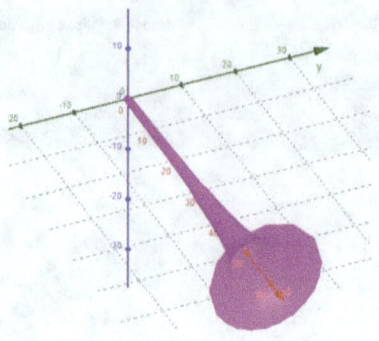
Figure 12 - $f_2(x)$ modelled in 3D.

Surface Area of a revolution

In this example, a student found the surface area of a Hershe's kiss to determine how much foil is wasted by wrapping them individually.

Hypothesis Testing (better for Applications and Interpretation)

This example is taken from a lesson on my website but is easily converted to an IA. Find out how many students study each science along with their gender. Then run a Chi Squared test of independence.

	Biology	Chemistry	Physics	Total
Girls	50	20	10	80
Boys	27	32	16	75
Total	77	52	26	155

Log-Log and Semi-Log Graphs

In this example, a student modelled the number of views of a viral music video against time and then linearised the data to help determine the relationship.

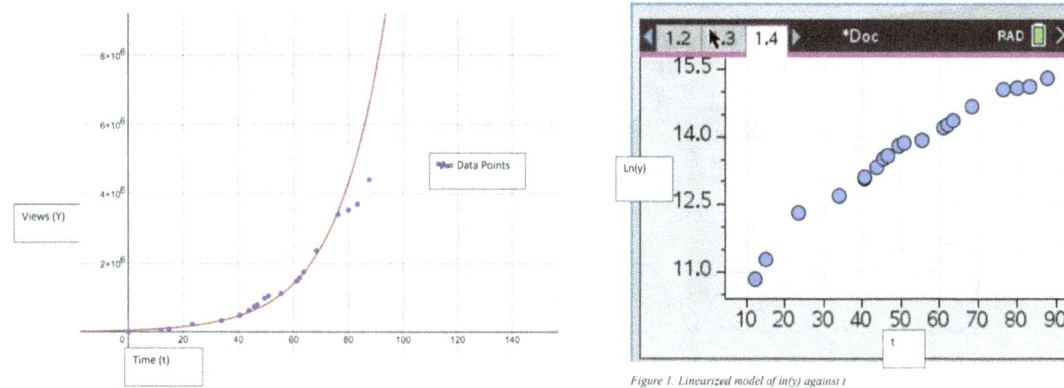

Figure 1. Linearized model of ln(y) against t

Optimization

In this example, a student used optimization to minimise the amount of plastic used in a plastic bottle.

Surface Area of a Truncated Cone = $\pi (r_1 + r_2) (S)$

Where: r_1 = Radius of the top of the Truncated Cone
r_2 = Radius of the base of the Truncated Cone
H = Height of the Truncated Cone
S = Slant Height of Truncated Cone: $\sqrt{(r_2 - r_1)^2 + H^2}$

With the above-mentioned formula in mind, I measured the required dimensions of the Kilimanjaro bottle by myself.

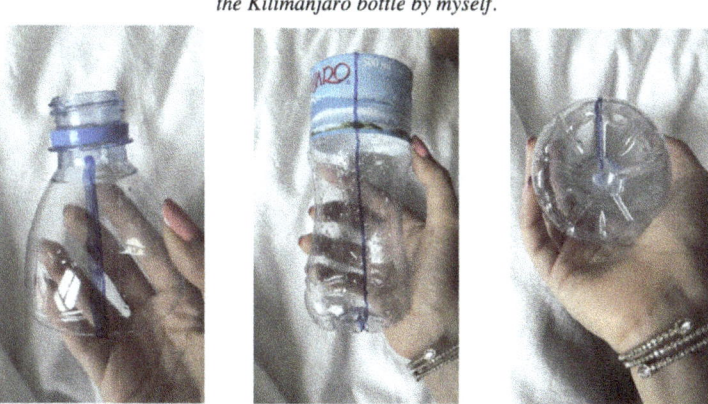

Figure 3: Measuring the dimensions of the 500ml Kilimanjaro Water Bottle

Kinematics

This example is taken from my YouTube video (IB Math IA: Hamilton v Verstappen). I modelled both Hamilton and Verstappen's velocity around a particular turn to compare.

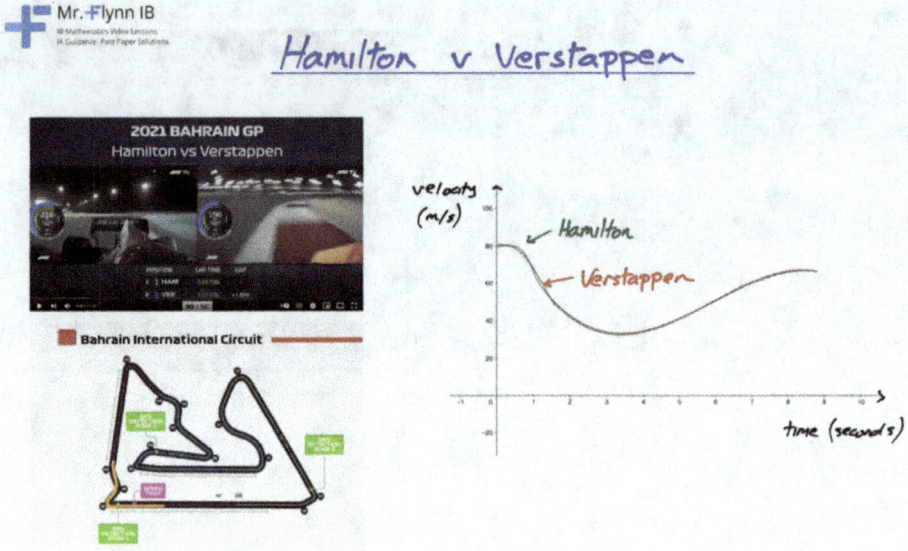

Voronoi Diagrams

This example was also taken from my YouTube video (IB Math IA: Voronoi Diagrams). I created a Voronoi diagram to determine the best place to open a café in Melbourne.

Graph Theory (better for Applications and Interpretation HL)

Another one taken from my YouTube video (IB Math IA: Graph Theory). The goal is to find the best route to take to visit a number of European cities.

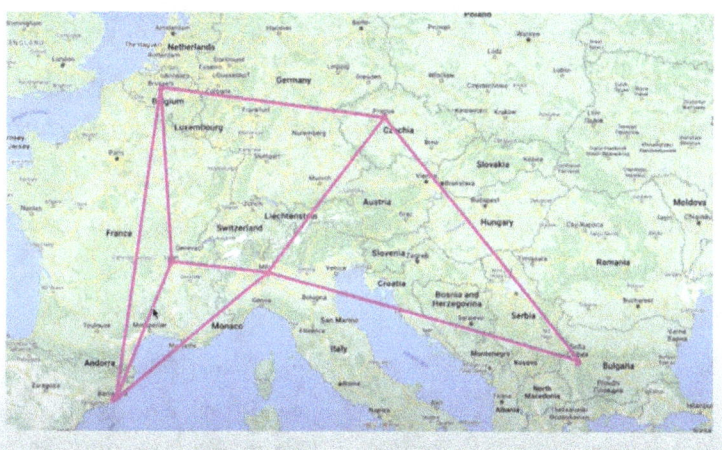

Topic Ideas to Fire up your Imagination

Here's a list of ideas I've come across or thought about over the years. Some are popular, some are unusual, and some are just weird. You don't need to use any of them exactly as written, but they're here to get your mind rolling and help you spot something that clicks.

Use them as starting points. The goal isn't to find a perfect idea — it's to find something you're curious enough to explore mathematically.

Modelling

- Time v anything
- TikTok Social Media Followers
- GDP
- Life expectancy
- Latitude and temperature
- Population
- Murder rates
- Prevalence of diabetes over time
- Gun violence
- Sleep
- Volume of ice caps
- Modelling a ski slope
- Plastic in the ocean
- GDP v environmental damage
- Carbon emissions
- Temperature v running times
- Age v running times
- Sunburn v time of day
- Drugs in sport
- Climate change
- Meat consumption
- Half-life exponential modeling
- Poverty
- Modelling tides

- Oil price
- Intelligence data
- Birth rates
- Literacy rates
- Social media (time spent)
- Social media (number of users)
- Corona virus/disease modelling
- Gini coefficient
- Refugee numbers
- Plastic in the ocean
- Happiness index
- Ferris wheel
- Sales
- Movies/Tv shows/Gaming
- Unemployment rates
- Team attendance
- When will India's population surpass China's?
- Relationship between running distances and times
- Height of water pouring into a glass
- Modelling tides to decide when to build a sandcastle

- BMI
- Tax rates v revenue - Laffer Curve
- Corruption index
- Air quality index
- Tennis height v aces
- Healthcare (patients per doctor…)
- 100m record progression
- Age v memory
- Cycling cadence v time
- Crime rates
- Alcohol/drugs
- Calories burnt
- Stock prices
- Coffee cooling
- Caffeine in blood
- Solar power
- Processor speed v time (Moore's Law)
- Sports teams wages v time or success
- Radioactive decay, when will it be safe to live in Chernobyl?
- Golf (distance v club)

Modelling a photo

- Modelling an arch
- Dubai fountains
- Apple logo
- Angry birds
- Skateboard Ramp
- Buildings
- High Heel Shoes
- A glass
- A dome
- Wavy hair
- Animals (e.g. a swan)

- Shape of Goggles
- Roller Coaster
- A lake or park from Google Maps
- Stadium
- Desert sand dunes
- Hot air balloon
- Food (e.g. fruit or croissant)
- Boomerang
- Eyes
- Tear drop
- A flower petal

- Racetrack corner
- Ski jump profile
- Modelling a bridge (quadratic v catenary)
- Musical instrument

- A shadow
- Sunglasses
- Mountain or landscape skyline or silhouette

Modelling a video

- Kitesurfing
- Basketball throw
- Jumping on a trampoline
- Horse jump
- Free kick
- Tennis shot
- Badminton
- Gymnastics
- Jump (skateboard, snowboard etc.)

- Modelling a swim stroke or dive
- Flip bottle challenge
- Boomerang flight
- Swing
- Drone
- Ice melting
- Skateboarder on half-pipe
- Fireworks

Statistics

- Teams' season
- Gender pay gap
- Police brutality
- Girls team v boys team basketball
- Messi v Ronaldo

- Fifa Stats
- Salaries
- Mo Farah v Usain Bolt
- Jordan v James
- Poisson Distribution for goals scored

Chi Squared test (goodness of fit)

- Birth months
- Colours of M & M s
- Spotify shuffle feature. Are songs played equally?
- Time/day people post in Instagram

- Days and detentions
- Voter turnout v age
- Does data follow a normal distribution?

Chi Squared test (test for independence)

- Gender
- Race
- Age
- Nationality
- Subject choice
- Favourite movie
- Sport
- Choice of social media
- Pet v exercise
- Homework submission v teacher
- Breakfast choice v wake-up time
- Strictness of teacher v subject
- Number of IB points

Normal distribution

- Times (swimming, running etc.)
- Distance (shotput, long jump etc.)
- Time on social media
- Include binomial distribution
- Probability of getting within 10cm of corner on a tennis court
- Basketball points scored in a season
- Change in stock prices
- Reaction time
- Time taken to solve a puzzle
- Time taken to read a page
- Distribution of pulse rate before and after activity

Area

- Area of a windsurf sail
- Area of a building
- Area of the Palm Dubai
- Area of eyes
- Pharmacokinetic curve
- Modeling skateboard jump. Which is the biggest jump?
- High jump v long jump which is bigger?
- Amount of sugar consumed by integrating a blood glucose concentration curve over time.
- Gini Coefficient
- Guitar/Violin/Musical Instrument
- Electrocardiogram waves
- Area of a logo
- Area under a bridge
- Heart rate v time. Area to give estimate of total cardiac workload
- Area of handwriting (compare letters)
- Total sound intensity exposure during a school day based on a decibel vs. time graph.

Volume

- Traditional pots/vases
- Which gold ornament is bigger?
- Volume of a glass
- Coffee pods
- Plane engine
- Doughnut
- Hot air balloon
- Building
- Beauty blender sponge
- Russian dolls
- Hershey drops
- Chess piece
- How many cream eggs could fit into an Easter egg?
- Soccer ball v rugby ball. Which is bigger?
- Hourglass
- Power plant
- Singapore's Super Trees
- Kuwait Towers
- Gherkin London
- F1 Wheel
- Perfume Bottle
- Burj Al Arab
- Volume of a bell
- Trumpet/musical instrument
- Volume of a dome
- Compare rockets
- Water capacity of an Arabic coffee pot
- Protein Powder v volume of container

Optimization

- Best seat at the cinema
- Minimising plastic
- Best cycling cadence
- Psychology memory experiment
- Best swim stroke rate
- Minimize surface area of an object while keeping volume fixed
- Best Temperature for running
- Tennis accuracy v power
- Long jump distance v run up
- Coffee v reaction times
- Surfing: best wave height

Kinematics

- Modelling parachute jump
- Who is a better driver?
- Roller coaster
- Modelling a bungee jump
- Modelling F1 driver. Velocity time graph.
- Dog speed v Time
- Elevator movement
- Fuel consumption rate curves
- 100m race
- Bike coming to a stop

Voronoi Diagrams

- Where to open a café?
- Nearest airport when flying
- Nearest hospital
- Where to live if I want to go to a particular school?
- Climbing Wall. Closest hold?
- Garden Sprinkler
- Emergency services in a city
- Fast food chains
- Voting areas
- Recreate a Voronoi on an animal (like giraffe)
- WIFI coverage
- Delivery Zones for restaurants

Graph Theory

- Best route while doing a campervan trip in Europe
- Mix Voronoi and Graph Theory
- Best school bus route
- Delivery routes for a delivery App

Log Graphs

- Modelling the price of Bitcoin
- Modelling the price of Nvidia
- Modelling number of views of a viral video
- Newton's law of cooling
- Population growth
- Spread of a virus
- Earthquake magnitude v energy released

Random thoughts

- Modelling a slinky going downstairs
- Normal distribution using Maclaurin series to estimate
- Find the upper limit that gives half the volume using Volume of a revolution
- Light reflecting on a curve
- Length of a curve
- Designing a cake
- Designing a logo
- Create a probability density function from data

Topics to Avoid

Here are some types of topics that are best avoided — they often lead to weak aims, limited maths, or very low engagement.

- Boring linear correlation (e.g. height v weight)
- Birthday paradox
- Golden Ratio
- Card games
- Chess
- Investigating Pythagoras' Theorem
- Casino games
- Fourier Series
- SIR models

A 5-Step Process for Choosing Your Topic

1. **Brainstorm 3–5 areas of personal interest.**
   Think beyond school — what do you *actually* care about?
2. **Research how maths connects to those interests.**
   Use YouTube, Wolfram Alpha, Reddit, and even ChatGPT.
3. **Narrow it down to something specific and investigable.**
   Can you ask a clear, focused question about it?
4. **Check the math level.**
   Will it include real mathematical analysis, not just data collection?
5. **Test the idea with a teacher.**
   A 5-minute conversation could save you weeks of wasted time.

What's Next?

By now, you should have a few topic ideas that genuinely interest you or at least a sense of where to look. Don't worry if you haven't chosen one yet. The goal is to start narrowing your focus.

In chapter 3, we'll take your topic and shape it into a clear, specific **aim**. This is where many students go wrong but if you get the aim right, the rest of your IA becomes much easier.

Chapter 3: The Aim

Your aim is **the most important sentence in your entire IA**. It's the foundation of everything else — your structure, your mathematics, your reflections, and ultimately, your mark.

Every part of your IA must relate directly back to your aim. If it doesn't, your work risks becoming unfocused, and you'll lose marks across all five criteria. Here's how:

- **Criterion A (Presentation)**: You can't score 4/4 if your IA includes irrelevant work — it won't be concise or coherent.
- **Criterion B (Mathematical Communication)**: If the maths doesn't support your aim, it's not considered relevant communication.
- **Criterion C (Personal Engagement)**: You can't fully engage with your IA if you aren't clearly exploring your own question.
- **Criterion D (Reflection)**: Your reflections won't be meaningful if they aren't tied to your original goal.
- **Criterion E (Use of Mathematics)**: Even strong maths can't score 6/6 if it isn't directly connected to your aim.

In short: if your aim is weak or unclear, your entire IA suffers even if your maths is good.

In this chapter, we'll focus on how to write a strong, focused, and relevant aim that sets your IA up for success.

Characteristics of a Strong Aim

A good aim is:

- **Clear**: The reader immediately knows what your exploration is about.
- **Focused**: Not too broad or vague. What exactly are you trying to do?
- **Mathematical**: It's obvious that math will play a central role.
- **Personal**: Shows some connection to your interests or experiences.

Tips for a good aim:

- Try to make your aim quirky or interesting. For example, instead of **"to calculate the area of the front of the school"**, try:
  "to determine how much paint is needed to paint the front of the school by modelling the school and calculating the area using integration."
- Ask yourself: "If I have a graph, what will the axes be?"

Weak v Strong Aim

Weak Aim	Strong Aim
Investigate the price of bitcoin	To determine the price of bitcoin in 2030 by mathematically modelling the price of bitcoin
To model a basketball shot	To compare my basketball shot to Steph Curry's by creating a mathematical model for both shots
To investigate the gender pay gap	To determine if there is a significant difference between pay for males and females in New Zealand
To model the Apple Logo	To calculate the proportion of a Mac cover that is taken up by the Apple Logo by creating a model and calculating the area using integration
To analyse my swimming times	To calculate the probability I win the 100m butterfly in the Dubai Open using Normal Distribution
Create a Voronoi diagram for Cafés in Melbourne	To determine the best location to open a Café in Melbourne by creating a Voronoi Diagram
To create a model of a chess piece	To calculate the volume of my silver Queen Chess Piece to determine its value

Template Phrases for Writing an Aim

I tell all my students to include the line **"the aim of this exploration is to..."**. This line should be in the Introduction. I also suggest writing this line in bold.

Do not assume the reader knows your aim.

Avoid phrases like "in this IA, I will talk about..." or "This is an IA on..."

And **never** use the term "research question". This is for an EE, not the Maths IA.

Reflecting on Your Aim

After writing your aim, ask yourself:

- Does this sound like a mathematics investigation?
- Can I explain *why* I'm doing this?
- Will the reader know what to expect?
- Will I find this interesting?
- Do I have a plan of what to do?

If your answer is yes to all these questions, then it is time to speak with your teacher.

Talk to your teacher

Remember your teacher is marking your IA, not the IB. It is vitally important to get them onside. You could ask your teacher questions like:

- Is my aim clear and specific?
- How could I make it stronger?
- Does this have the potential to achieve a high grade?
- Is it realistic to explore fully in 12–20 pages?
- Would you be interested in reading this?

If they don't like the idea, follow up with:

- Why is it not a good aim and how can I improve it?
- Can you suggest a related aim that would work better?
- What kinds of aims have you seen in the past that scored highly?

If your teacher still isn't happy with your aim, then it's probably time to change topic.

If your teacher is happy with your aim, then you are ready to **do the math**.

What's Next

That's where we're going next. Now that you've got a solid aim — one that's focused, mathematical, and approved by your teacher, it's time to bring it to life. In the next chapter, we'll dive into how to choose the right mathematical tools, apply them effectively, and make sure the maths in your IA is both meaningful and impressive.

Chapter 4: Do The Math

Now that you have a solid aim, it is time to **do the math**. A common mistake I see students make is that they try and write their IA before doing the maths. I advise doing all the maths first. You can even do it on paper. It takes of lot of time and tedious work to write a mathematical exploration on a word processor. Make sure you have a solid IA before doing this.

This chapter is about choosing the right **mathematical tools**, applying them correctly, and making sure they're deep enough to satisfy the examiners. This is where your IA becomes more than just a school project, it becomes a piece of real mathematical work.

What Kind of Math Should I Use?

This depends on your level and your topic, but here are some rules:

Rule	Why It Matters
The maths must be central.	Your IA is not a science or economics paper. The maths should drive the investigation.
It must be appropriate to your level.	The Maths used should be at the level of the syllabus or just beyond. Prior learning only won't suffice.
You must understand it.	Don't include maths you don't understand. This will be obvious to your teacher and the examiner. You will lose marks in all 5 criteria.

How Much Maths Is Enough?

The IB doesn't specify a set number of equations or techniques. Instead, they look for:

- **Correctness:** The maths you use is accurate and well-explained.
- **Sophistication:** It shows a level of understanding that matches your course level.
- **Purposefulness:** The maths actually helps you move towards your aim.
- **Clarity:** You explain *what* you're doing and *why* — not just showing calculations.

Common Mistakes to Avoid

- **Including too much maths**: Some students think "the more, the merrier," but adding extra, irrelevant mathematics will actually cost you marks. Examiners want focus, not quantity.
- **Using overly complex math you don't fully understand**: Thinking "the more complicated, the better" is a trap. If you can't explain it clearly, you'll lose marks even if the math looks impressive.

Showing the Process

Your IA should show the mathematical process, not just the final result. That means:

- Stating formulas before using them
- Showing solving equations step-by-step
- Including diagrams, graphs, or tables where appropriate
- Explaining what each step is doing and why

Understanding vs. Overusing Technology

Using tools like Desmos, GeoGebra, Excel, or Python is encouraged but you must still understand what's happening.

- If you generate a regression equation using a calculator, explain what it means.
- If you create a model using GeoGebra, you must explain why you chose that model.
- If you run a simulation in Python, show and interpret the output.
- Don't copy-paste computer results, explain them in your own words.

Pro tip: "Calculator-generated maths" should support your IA, not replace your own thinking.

Using Real Data

If your IA involves data:

- Use reliable sources (cited properly).

- Show how you processed or transformed the data.
- Use statistical tools meaningfully.
- Comment on accuracy, outliers, or limitations.

Toolbox of Common Techniques

Check out my YouTube channel for a detailed explanation on how to apply several techniques like fitting functions onto a picture or set of data.

Pro Tip: Ask your teacher for help. Your teacher is only allowed to give formal feedback on one full draft, but they're encouraged to support you throughout the entire IA process. Don't be afraid to ask questions, check your understanding, or get guidance along the way.

What's Next?

Once you've done the math, it's time to learn more about **The Criteria**. What does it mean to score well in "Presentation"? What does "Personal Engagement" *actually* mean?

That's what we'll break down in Chapter 5: *Marking Criteria Breakdown*.

Chapter 5: The Criteria

What are examiners actually looking for?

The IB Maths IA is marked out of **20 marks**, split across five criteria. Understanding how these work is crucial because many students lose marks not from poor math, but from failing to demonstrate things examiners are actively looking for. Let's break down each criterion in student-friendly terms and show you what a 7-worthy IA looks like in each one.

A: Presentation

IB Descriptor

Level	Descriptor
0	The exploration does not reach the standard described by the descriptors below.
1	The exploration has some coherence or some organization.
2	The exploration has some coherence and shows some organisation.
3	The exploration is coherent and well organized.
4	The exploration is coherent, well organized, concise.

What do I need to get 4/4?

- A Front page that has a title and number of pages.
- It should be 12-20 pages with double line spacing excluding bibliography.
- You should have a Bibliography (if needed).
- Your pages are numbered.
- An Introduction which includes an aim, rationale and plan.
- A clear and focused aim.
- A conclusion which refers back to your aim.
- Everything is relevant to your aim.
- It should be Easy to follow (for a student in your class).
- It flows from one section to the next.
- Graphs, tables and diagrams are in the appropriate place. I.e. I don't have to go searching for a graph that you are describing.
- Graphs, tables and diagrams have proper lead ins and follow ups.
- Big tables of data can go in the appendices.

- Every page is important (concise IA).
- Looks Professional.
- Microsoft Word is better than Google Docs.
- No repetitive calculations.
- Correct citations.

B: Mathematical Communication

IB Descriptor

Level	Descriptor
0	The exploration does not reach the standard described by the descriptors below.
1	The exploration contains some relevant mathematical communication, which is partially appropriate.
2	The exploration contains some relevant appropriate mathematical communication.
3	The mathematical communication is relevant appropriate and mostly consistent.
4	The mathematical communication is relevant, appropriate and consistent throughout.

What do I need to get 4/4?

- Define all key terms and variables (don't use x, if you haven't defined what it is).
- Define key terms and variables when they first appear. Not at the beginning of the IA.
- Mathematical language is correct (notations, symbols and terminology). E.g. Don't use * for a multiplication sign.
- Variables are italicized.
- Functions are **not** italicized e.g. $\sin(x)$ **not** $sin(x)$.
- Use equation editor. $2x + 1 = 7$ looks better than 2x + 1 = 7.
- Put equations in centre of page. Explanations go on the left.
- Diagrams, tables, charts, graphs and models are included and look good and professional. Looks matter a lot in an IA. Generally put in centre of page.
- Make sure all diagrams, tables, charts, graphs and models have titles and that axes (when they exist) are labelled.
- Use appropriate degree of accuracy and explain why. E.g. My highest jump was 1.35446336653334 metres does not look good. 'High jump records are

measured to the nearest cm; therefore, I will be rounding all my answers to the nearest cm or 2 decimal places. My highest jump was 1.35 metres.' This looks much better.
- Graphs and diagrams should have proper lead ins and follow ups leaving clear purpose.
- Approximation symbol should be used when you have rounded.
- No calculator screenshots.
- Equations should be on one line.
- Tables shouldn't go over two pages.

C: Personal Engagement

IB Descriptor

Level	Descriptor
0	The exploration does not reach the standard described by the descriptors below.
1	There is evidence of some personal engagement.
2	There is evidence of significant personal engagement.
3	There is evidence of outstanding personal engagement.

What do I need to get 3/3?

- You need to care about your exploration and your results. This needs to be evident.
- You need to come up with your own ideas and be creative.
- Explain the purpose of the processes used to achieve the aim.
- The IA should be personal. Finding the probability of making the track team is more personal than investigating Pythagoras' theorem.
- You could do research and collect secondary data.
- You could collect your own data.
- You could do a questionnaire.
- You could create your own model from a video or photo.
- You could do an experiment.
- You could learn new maths.
- You could learn how to use new software like GeoGebra or Logger Pro.
- You could refine a model.
- Give your opinion whenever possible.
- At least some of the time, write in the first person.

- Look at real world situations.
- Consider different perspectives (historical, global, local).
- Your IA should be original and must be unique.
- Passion and interest should be abundant in the overall read of the paper.
- This is about engagement with the mathematics.
- Personal engagement should drive the exploration forward.
- Consider alternative approaches, test predictions, and present findings in creative ways.
- Explore topics from multiple perspectives and think critically about your methodologies.
- Show creative and independent thinking in reacting to problems as they arise.

D: Reflection

IB Descriptor

Level	Descriptor
0	The exploration does not reach the standard described by the descriptors below.
1	There is evidence of limited reflection.
2	There is evidence of meaningful reflection.
3	There is evidence of critical reflection.

What do I need to get 3/3?

- You should have reflection in the conclusion but not **only** in the conclusion. It should be throughout.
- Reflect in the first person.
- Comment on **all** your results.
- Discuss what your results mean in context. E.g. '$r = 0.8$ means there is strong positive correlation' is not good enough. '$r = 0.8$ indicates strong positive correlation which suggests that the hotter it gets, the more ice-cream I will sell is discussing in context.
- Critically evaluate your results. E.g. $r = 0.8$ suggests that hotter means I sell more ice-cream but where does it end? If it is 48 degrees, will anyone even come to my ice-cream stand?
- Have you discussed limitations. E.g. $r = 0.8$ but I only collected data over 15 days. Is this enough? Etc.

- State your assumptions. E.g. The probability of scoring a penalty remains constant.
- Discuss issues you had when writing the IA. E.g. I wanted to collect data over 6 months, but we only had 2 months to do the IA.
- Discuss what you discovered, what you found interesting, difficult, what did you learn etc.
- What were the strengths and weaknesses?
- Discuss what you could have done to make the IA better.
- Discuss alternative mathematical approaches.
- Discuss possible extensions.
- Critical reflection should drive the next steps.
- Consider how the chosen mathematical methods impacted the overall conclusions.
- Consider how the results affected the direction of the investigation.
- Reflect on the results of each section before moving to the next.
- Are your analyses and strategies and results appropriate?
- If creating a model, test it. Think about does it make sense and comment.
- Think how your method can be improved.
- Is your level of accuracy appropriate?
- How confident are you in your results? Can you quantify this?
- Did you consider alternative approaches?
- Reflection is better in full sentences/paragraphs rather than a bullet list or table.
- Reflection should create questions which then guide the exploration forward.

E: Use of Mathematics

IB Descriptor (SL Only)

Level	Descriptor
0	The exploration does not reach the standard described by the descriptors below.
1	Some relevant mathematics is used.
2	Some relevant mathematics is used. Limited understanding is demonstrated.
3	Relevant mathematics commensurate with the level of the course is used. Limited understanding is demonstrated.

4	Relevant mathematics commensurate with the level of the course is used. The mathematics explored is partially correct. Some knowledge and understanding are demonstrated.
5	Relevant mathematics commensurate with the level of the course is used. The mathematics explored is mostly correct. Good knowledge and understanding are demonstrated.
6	Relevant mathematics commensurate with the level of the course is used. The mathematics explored is correct. Thorough knowledge and understanding are demonstrated.

IB Descriptor (HL Only)

Level	Descriptor
0	The exploration does not reach the standard described by the descriptors below.
1	Some relevant mathematics is used. Limited understanding is demonstrated.
2	Some relevant mathematics is used. The mathematics explored is partially correct. Some knowledge and understanding are demonstrated.
3	Relevant mathematics commensurate with the level of the course is used. The mathematics explored is correct. Some knowledge and understanding are demonstrated.
4	Relevant mathematics commensurate with the level of the course is used. The mathematics explored is correct. Good knowledge and understanding are demonstrated.
5	Relevant mathematics commensurate with the level of the course is used. The mathematics explored is correct and demonstrates sophistication or rigour. Thorough knowledge and understanding are demonstrated.
6	Relevant mathematics commensurate with the level of the course is used. The mathematics explored is precise and demonstrates sophistication and rigour. Thorough knowledge and understanding are demonstrated.

What do I need to get 6/6?

SL and HL

- Make sure the mathematics is not just prior learning like Pythagoras or SOH CAH TOA.

- Make sure you understand all the mathematics. If you don't, it will be obvious.
- Clearly explain each step of your working.
- Make sure the mathematics is correct.
- All the mathematics should be relevant. E.g. (of what not to do) r = 0.2 I will now find the regression equation to make predictions.
- If you use technology to find a solution, make sure you have demonstrated understanding.
- Don't use very complicated mathematics if simpler mathematics could have been used.
- Needs to be understandable to a peer.
- Consider the behaviour of a model before fitting a model using technology.
- Your reflection will help to demonstrate understanding.
- All maths should be relevant to your aim.

HL only

- 'Sophistication' means commensurate with the HL course or, if found in the SL course it is beyond what an SL student could reasonably be expected to do.
- 'Precise' means error free.
- 'Rigour' involves clarity of logic and language when making mathematical arguments and calculations.

What's next?

Now that you understand how the IA is marked, you know exactly what examiners are looking for and where most students lose marks.

In the next chapter, we'll go even deeper, drawing on insights straight from the Principal Examiner Reports. These are full of honest advice, recurring mistakes, and behind-the-scenes guidance that can make a real difference to your final mark.

40

Chapter 6: Tips from the Principal Examiner

Every year, the IB publishes examiner reports that highlight what students are doing well and what they're consistently getting wrong. These reports come straight from the people who mark your IA, including the Principal Examiner, and they're full of valuable insights.

I've read them all.

In this chapter, I've pulled together the most important, most repeated, and most useful advice from those reports. These are the things the examiners wish every student knew before submitting their IA — the things that can push you up a grade, or quietly pull you down if ignored. Take these tips seriously. They come from the very people who decide your marks.

General Recommendations

- **Choose Topics Wisely**: Select a topic of personal interest that aligns with the syllabus and allows for creative mathematical exploration. Avoid overused topics (e.g., Birthday Paradox, SIR model, Monty Hall problem, Golden ratio), or research-style reports that rely on sourced work, as these limit originality and engagement.
- **Understand Assessment Criteria Early**: Review the IA criteria and level descriptors before starting. Discuss sample explorations in class to understand expectations and common pitfalls.
- **Seek Teacher Guidance**: Work with teachers to refine topic choices and receive feedback on one draft.
- **Cite Sources Properly**: Cite all sources at the point of use and in a bibliography, ensuring accuracy to avoid plagiarism. Reference AI tools clearly if used, and avoid over-reliance that diminishes your voice.
- **Use Technology Appropriately**: Leverage graphing calculators or software for calculations to focus on analysis and reflection. Avoid including GDC instructions or screenshots, as they do not enhance mathematical communication.
- **Proofread and Format Correctly**: Double-space the exploration, use a minimum font size of 11, number pages, and include only the title and page count on the cover page. Convert to PDF before uploading to ensure mathematical expressions and diagrams are clear and correctly oriented. Stick to the recommended 12–20 pages with double spacing.
- **Too many Linear Correlation IAs**: Maybe do a different topic or if you really want to do linear correlation, make it stand out from the crowd.

Criterion A: Organization and Coherence

- **Craft a Clear and Focused Aim**: State a concise, achievable aim in the introduction to guide the exploration. Avoid overly ambitious or multilayered aims that lead to incoherence or lack of concision. Ensure the aim is precise to facilitate a coherent and complete exploration.
- **Maintain Relevance to the Aim**: Ensure all sections and mathematical processes directly support the stated aim. Avoid irrelevant extensions, repetitive content, or sections unrelated to the central theme, as these detract from coherence and may lower marks.
- **Prioritize Conciseness**: Keep the exploration concise by avoiding long-winded introductions, unnecessary rationales, research questions, or textbook-style explanations of syllabus content (e.g., calculating a mean). Stick to the recommended 12–20 pages with double spacing, excluding appendices, to maintain focus.
- **Avoid Unnecessary Structural Elements**: Exclude tables of contents, detailed methodology plans, research questions, or sections labelled "evaluation," "limitations," or "extensions" at the end, as these disrupt flow and repeat content. Integrate explanations of processes as they occur, and avoid subheadings that signal specific criteria (e.g., personal engagement).
- **Ensure Readability for Peers**: Write for an audience of classmates familiar with the syllabus. Clearly explain non-syllabus concepts or terminology from other disciplines, but avoid over-explaining syllabus content to maintain concision.
- **Structure the Conclusion Effectively**: Summarize key findings succinctly in the conclusion without introducing new analysis. Avoid lengthy post-conclusion discussions to maintain organization.
- **Balance Concision and Completeness**: Present essential calculations and results (e.g., observed/expected values for statistical tests) in the main body, not relegated to appendices. Large data tables or extensive calculations should be in appendices but referenced appropriately.
- **Number Pages**: Ensure all pages are numbered for clarity and ease of navigation.
- **Introduce Visuals Clearly**: Introduce tables, graphs, and diagrams with clear explanations in the text to connect them to the exploration's narrative and aim.

Criterion B: Mathematical Communication

- **Use Correct Notation**: Avoid calculator/computer notation (e.g., "*" for multiplication, "^" for exponents, "/" for fractions, "Normcdf" outputs). Use proper mathematical symbols (e.g., "≈" for approximations, subscripts like Q_1 instead of Q1) and italicize variables/parameters in both equations and text.
- **Label and Format Clearly**: Ensure graphs, tables, and diagrams are properly labelled with appropriate scales, units, and context. Label axes even when zooming in on graphs, and avoid splitting tables or equations across pages.
- **Maintain Consistency**: Use consistent notation (e.g., avoid mixing upper/lowercase letters for the same variable) and appropriate, justified levels of accuracy in context, not default rules like 3 significant figures. Discuss the degree of accuracy in relation to the exploration's context.
- **Use Equation Editors**: Present mathematical expressions using an equation editor for clarity. Avoid screenshots of calculator outputs, as they include irrelevant data and do not enhance communication.
- **Incorporate Relevant Visuals**: Include student-created or properly referenced diagrams and graphs that support the analysis. Avoid irrelevant visuals or long tables in the main body; place extensive tables in appendices with key excerpts in the text.
- **Define Key Terms at Point of Use**: Define variables and non-syllabus terminology (e.g., statistical terms or concepts from other disciplines) in context when they first appear, not in a separate list at the beginning.
- **Maximize Achievement with Feedback**: Use teacher feedback on drafts to eliminate careless mistakes (e.g., unlabelled diagrams, incorrect notation) and aim for the highest level as it is achievable in this criterion with attention to detail.

Criterion C: Personal Engagement

- **Engage with Mathematics, Not Just Context**: Demonstrate engagement through creative and independent mathematical approaches, not just personal interest or effort in the topic. Show your own ideas by exploring the topic from multiple perspectives, testing predictions, or refining models.
- **Drive Exploration with Mathematical Purpose**: Explain the purpose of mathematical processes used to achieve the aim.
- **Avoid Template-Based or Generic Work**: Steer clear of formulaic explorations (e.g., correlation-based IAs in AISL, Birthday Paradox, SIR model, Monty Hall problem, Golden ratio, Torricelli Trumpet paradox), as these limit creativity and

personal ownership. Topics easily found online or in textbooks often score poorly due to lack of originality.
- **Leverage Personal Experiments**: Designing experiments, surveys, or questionnaires can enhance engagement if results drive the mathematical analysis. However, data collection alone is insufficient without creative application.
- **Avoid Over-Reliance on Published Sources**: Explorations based on sourced work or textbook proofs restrict opportunities for independent thinking and creativity, limiting scores. If you can Google it, it's probably not a good IA.
- **Show Authentic Problem-Solving**: Demonstrate authentic engagement by adapting to challenges as they arise in the exploration, rather than following a pre-set path of standard techniques.

Criterion D: Reflection

- **Engage in Ongoing Critical Reflection**: Reflect throughout the exploration, not just in a final paragraph. Analyse the appropriateness of methods, correctness of results, and their impact on the aim to guide the exploration's progress. Use full sentences, not bullet points or tables.
- **Ask Critical Questions**: Pose questions like "How reliable is this result?", "What if...?", or "How can the method be improved?" to drive meaningful reflection. Consider alternative approaches and evaluate their viability.
- **Avoid Descriptive Summaries**: Move beyond describing results or listing generic limitations/extensions, as these only merit low marks. Reflections must be evidence-based and justified, not speculative or based on personal assumptions.
- **Contextualize Reflections**: In statistics-based explorations, reflect on data validity, sampling methods, sample size, and the appropriateness of statistical tests. In modelling, evaluate the model's suitability for the phenomenon, including domain/range and impact of accuracy/rounding.
- **Test Model Viability**: For modelling explorations, interrogate the model's real-world applicability and test its viability. Justify the choice of functions by analysing data shape and trends, not using guess-and-check or high R^2 values alone.
- **Assess Accuracy and Confidence**: Discuss the confidence in results, especially when methods introduce uncertainty. Quantify potential errors and justify the chosen level of accuracy in context.

Criterion E: Use of Mathematics

- **Ensure Relevance**: Use mathematics directly relevant to the aim. Irrelevant or overly complex methods (e.g., unnecessary calculus, misapplied statistical tests) will not earn high marks, even if correct.
- **Demonstrate Understanding**: Explain mathematical processes in your own words, justifying their relevance and interpreting results in context. Simply performing calculations or quoting formulas without explanation does not show understanding.
- **Stay Within Syllabus Scope**: For HL, use mathematics at or slightly beyond the syllabus to demonstrate sophistication, but ensure clear explanations. Avoid overly advanced methods (e.g., Fourier/Laplace Transforms). Mathematics from the SL syllabus can achieve high marks in HL if used in a complex way beyond SL expectations.
- **Follow Modelling Processes**: In modelling explorations, plot data first, analyse trends, and justify model choices based on the phenomenon's properties, not just high R^2 values or guess-and-check methods. Consider the function's behaviour before/after the data domain. R^2 is suitable for interpolation (e.g., fitting a curve to a vase) but not extrapolation (e.g. predicting the population in 40 years).
- **Avoid Manual Calculations for Syllabus Methods**: Do not calculate syllabus-based formulas (e.g., Pearson's correlation coefficient) by hand with long tables, as this shows no more understanding than using a GDC and impacts conciseness negatively.
- **Explain Non-Syllabus Mathematics**: Clearly explain any mathematics beyond the syllabus (e.g., Surface area of a revolution) to ensure a peer can understand without external sources.

What's next?

You've seen exactly what examiners look for and where most students slip up. Now it's time to put it all together.

In the next chapter, we'll walk through the writing process: how to structure your IA, how to present your ideas clearly, and how to make sure your final draft is focused, polished, and easy to read.

Chapter 7: The Writing Process

You've chosen your topic, crafted your aim, and done the math. Now comes the part that separates a 5 from a 7: **writing**.

The IA is not just about doing mathematics, it's about communicating it. If your IA is hard to follow, unclear, or full of unexplained steps, your marks in all 5 criteria will suffer — no matter how good your ideas are.

This chapter will show you how to structure your IA like a professional report, write in a mathematical yet readable tone, and present your work clearly and logically.

A Strong IA needs A Strong Structure

There's no official template, but most high-scoring IAs follow a structure something like this (you can change the headings, although always keep the Introduction and the Conclusion):

1. **Cover Page**

- This includes a title and number of pages.
- Give your IA a nice quirky title. Not something boring that nobody will want to read. For example: Instead of 'Calculating the volume of two hotels', why not 'Finding the GOAT hotel'?.

2. **Introduction**

- Introduce your topic.
- Explain your rationale. Why did you choose the topic?
- State your aim. "The aim of this exploration is…".
- Write a brief plan. What are you going to do?

3. **Data collection or Modelling**

- Introduce your data here. What is it? Where did you get it?
- If you are creating a model, you can do this here? Explain the process.

4. Mathematical Process

- This is the *core* of your IA.
- Introduce and apply mathematical tools step-by-step.
- Explain what each method does and why you're using it.
- Include graphs, diagrams, equations — and annotate them clearly.
- For example, if you are calculating the area under a curve, you can do this here.
- Avoid dumping calculator output with no explanation. Instead, guide the reader through your thought process.

5. Interpretation & Discussion

- This is your voice.
- What do your results mean?
- How do they help answer your aim?
- Is the model realistic? What are the limitations?
- How would different assumptions affect the outcome?
- What did you learn?
- Were there challenges or surprises?
- How did your understanding of the maths or the context grow?
- What would you do differently?
- These questions can and should also be answered in sections 2 and 3.

6. Conclusion

- Restate your aim in the conclusion and discuss if you reached your aim or not.
- Summarise what you found and how the maths helped.
- Don't introduce new math here, just tie things up neatly.

7. Bibliography / Citations

- IB requires full academic honesty.
- Cite all sources used.
- Use a consistent format (e.g., APA, MLA) and include URLs where appropriate.

8. Appendices

- Include any large tables of data that haven't made it into the main section.

Formatting & Visual Clarity

Make your IA easy to read. Use:

- Clear headings.
- Double Spacing.
- 12-20 pages. Stick to this.
- Pages numbered.
- **No** table of contents.
- Title all graphs and diagrams.
- Label all graphs.
- Proper notation (e.g. variables in italics).
- Keep plenty of Space between paragraphs, graphs and sections.

Writing Style Tips

- Use Microsoft Word. Not Google Docs. Equations look much better.
- Write in the first person.
- I recommend writing in the past tense (but it doesn't have to be).
- Keep language simple. No marks for fancy language. And sometimes negative marks (if you don't understand what you are saying).
- Avoid long blocks of text.
- Don't just describe — analyse.
- AI can help with proofreading but don't let it write the IA for you. Simple, personal language that is yours works best.

Flow State

A word that examiners often use is **"flow."** Your IA is not meant to read like a formal paper for a mathematical journal — it's a short exploration written by a high school student.

That means it should feel natural and easy to follow. Avoid having lots of disconnected sections. Instead, aim for a smooth progression from one idea to the next. In fact, you might even consider minimising the number of sections altogether, especially if your IA reads well as a continuous narrative.

What's next?

Your IA is written, your maths is done, and everything's in place. But before you submit, there's one more essential step: the final check.

In the next chapter, we'll go through a detailed checklist to help you catch small mistakes, tighten up your work, and make sure you haven't missed anything that could cost you marks.

Chapter 8: Final Checklist

Before you submit your IA, go through this checklist carefully. Tick each item once you're confident it's complete.

This is your last chance to catch small mistakes, make improvements, and ensure your IA is clear, focused, and ready to be marked.

Overall

- ☐ I have acted on my teacher's feedback
- ☐ I have not copied from another source without citation
- ☐ No sections were written by AI
- ☐ I fully understand my IA
- ☐ I have an interesting title
- ☐ I do **not** have a table of contents
- ☐ I have saved my IA in multiple locations
- ☐ Someone (other than my teacher) has read my IA for clarity and flow
- ☐ I am proud of the work I am submitting

Criterion A

- ☐ My IA is between 12 and 20 pages
- ☐ The cover page has a title and number of pages
- ☐ Pages are numbered
- ☐ I have checked spelling and grammar
- ☐ If needed, there is an appendix and bibliography
- ☐ The IA has an Introduction and Conclusion
- ☐ The Introduction has an aim, rationale and plan
- ☐ The aim is clear and focused
- ☐ The IA is logically structured with clear headings
- ☐ The IA is written in clear straight-forward language
- ☐ The IA flows from one section to the next
- ☐ Graphs, diagrams and tables have titles and axes are labelled
- ☐ Graphs, diagrams and tables are introduced
- ☐ A peer could read and understand the IA
- ☐ All sources are properly cited

Criterion B

- ☐ Mathematical notation is correct (e.g. I have **not** used * for multiply)
- ☐ I have used an equation editor (e.g. 3x+4=7 does not look good)
- ☐ All my variables are defined (when introduced) and italicised
- ☐ Equations are in the centre of the page, explanations on the left
- ☐ Graphs, diagrams and tables have titles and axes are labelled
- ☐ I have talked about degree of accuracy (rounding)
- ☐ If I have rounded, I have used the ≈ sign instead of the = sign

Criterion C

- ☐ I am genuinely interested in this IA/topic
- ☐ The IA feels like my work
- ☐ I have made the exploration my own
- ☐ I care about the IA and the results
- ☐ I have written at least some of the time in the first person
- ☐ If appropriate, I have included some photos
- ☐ I care about the maths and the maths drives the exploration
- ☐ I have given my opinion
- ☐ I have considered different perspectives
- ☐ My IA is unique
- ☐ I have considered alternative approaches
- ☐ I have thought critically about the methods I used
- ☐ I have thought creatively or independently
- ☐ My thinking and decision-making is evident throughout

Criterion D

- ☐ I have reflected throughout the IA and not just at the end
- ☐ I have reflected after each result
- ☐ I have reflected in the first person
- ☐ I have **critically** evaluated my methods and results
- ☐ I have considered limitations and weaknesses
- ☐ I have stated my assumptions
- ☐ I have discussed what I discovered and what I found interesting
- ☐ I have discussed alternative mathematical approaches
- ☐ My reflection helps makes decisions about where to go next

☐ If I created a model, I tried to refine it
☐ I have considered how confident I am with my results
☐ All reflection helps with my aim
☐ I have discussed possible improvement and extensions

Criterion E

☐ All mathematics is relevant to the aim
☐ The mathematics is commensurate with the level of the course
☐ The maths is correct
☐ If I have used a formula, I have explained what it means
☐ Mathematical steps are shown clearly
☐ If I have used technology, I have demonstrated understanding
☐ I understand and could explain every part of my IA
☐ A peer could understand my IA

You've now worked through the full IA process — from understanding the task to crafting a strong aim, doing meaningful mathematics, and writing with clarity and purpose. You know what the examiners are looking for, and you've seen how to meet the criteria step by step.

If you've already dipped into the sample IAs in Section 2, great — now you'll be able to view them with fresh eyes and a stronger sense of what makes them successful. If not, this is the perfect time to explore them properly. Use them for inspiration, comparison, or simply to reassure yourself that you're on the right track.

Section 1 has given you the map. Section 2 will show you the terrain. Let's take a closer look.

Section 2

13 IA Samples with Examiner Comments

The Internal Assessments in this section are recently submitted examples that achieved high marks after IBO moderation. To maintain academic integrity, appendices have been removed to prevent plagiarism or duplication of results. The IAs are reproduced exactly as originally submitted, with no changes to content or formatting, apart from the removal of page numbers to ensure consistency within this book. Copyright remains with the original authors, and this publication is not endorsed by the IBO. Each IA is reprinted here with the permission of its author.

IA Sample 1 Examiner Comments: Table Tennis Shot

This is a strong and personally engaging SL IA. The student mathematically models his forehand and backhand table tennis shots to compare.

Appropriate for: AASL, AISL

Possible for: AAHL, AIHL (Student may need more sophistication and rigor)

Criterion	SL	HL
A	4	4
B	3	3
C	3	3
D	3	3
E	5	4
Total	18	17

A: The IA is structured logically, progressing from data collection to modelling, comparison, and conclusion. The student includes relevant diagrams and screenshots that help the reader understand the context and data. The mathematical work is well-ordered and easy to follow, with equations and graphs clearly presented. The exploration is easy to follow and well written. It is well organised and concise.

B: Variables are defined, and the notation is mostly correct and consistent. Graphs are well presented and support the written explanation well. The student communicates the mathematics clearly. The use of Google Docs instead of MS Word takes from the mathematical presentation.

C: This is a clearly personal and original exploration: the student filmed and analysed their own table tennis shots. The decision to compare forehand and backhand shots shows initiative and genuine interest. The use of Logger Pro demonstrates active problem-solving and technological engagement. Engagement is evident in the effort to extract and compare meaningful mathematical quantities (angle, landing point, area).

D: The student reflects meaningfully on the mathematical results and their real-world implications. For example, they consider how the angle of release affects shot trajectory and landing, and they interpret the area under the curve in a physical context. They discuss the limitations of the model, such as assuming a smooth parabola and ignoring air resistance or spin. Lots of critical reflection seen throughout.

E: The mathematics used — fitting quadratic models, calculating area under the curve (via definite integrals or estimation), analysing angles and landing points is appropriate for SL level. All techniques are correctly applied and show a clear understanding. The student uses them with insight and applies them purposefully. They have used technology appropriately to support graphing and regression. 6/6 was possible here but more thorough explanations would have secured it.

Introduction

Mathematical analysis is of high importance for understanding how our world works. Through the study of quantity, one can discover certain aspects and patterns which may not be anticipated by the naked eye. Mathematical justification can offer quantitative reasoning to our doubts or unresolved matters. Within the scope of professional sports, mathematical analysis of current performance has allowed individuals to observe the flaws in their techniques through a numerical lense. This has enabled them to correct past mistakes and tap into new potential. Correlations of data can help resolve important problems and optimize techniques through mathematical proof.

My interest in table tennis arose at the age of nine when I started playing in the garden with my father. Ever since, it has always been a great passion of mine. I started training at a local club and enjoyed learning about the sport. At that age, I thought that table tennis was quite easy. However, as I watched professional players play, I realized that there was more to the game than just hitting a ball back and forth. In a game of table tennis, there are a wide range of shots you can hit. For example, topspin, block, smash, push and slice. However, what complicates the sport is the fact that each of the previously mentioned shots can be executed with different trajectories, spins, speeds and angles depending on how you hit the ball.

By the age of thirteen, I had won twelve national medals and placed 16th in the under-13 European championship. Whilst competing in different countries around the world, I realized that I was winning most of my rallys with my backhand. My backhand shots were more consistent and when I hit them, I was more likely to win the point. Although it may have been psychological, I was more confident with my backhand than my forehand and would aim to finish important rallies with this shot.

"Knowing yourself is the beginning of all wisdom" - Aristotle

Although I am very proud of my accomplishments, as a motivated sportsman, I always aim to become better and improve. In order to do so, I must investigate further into my game. Therefore, *the aim of this exploration is to model my forehand and backhand topspin shots in*

A+ Good Intro and clear aim

order to find out why more of my points are won with my backhand. Understanding what makes the shot more effective will help me further invest in training on that specific area and look for ways to evolve my forehand to reach the same level of effectiveness.

Collecting data

Throughout this exploration, mathematics is used in forms of quadratics and integration in order to justify my qualitative observations. Moreover, several mathematical websites (Logger pro and Desmos) will be used in order to show trends through graphs and diagrams.

To gather my data, I asked my sister, a former national champion, to block my forehand and backhand topspin. Whilst we were playing, a slow-mo camera, used by my mother, recorded each shot at the height of the table tennis table. After hitting a forehand and a backhand topspin successfully, the rest of the data was collected electronically leveraging the use of two applications.

An application named 'Logger Pro' allowed me to upload the video taken and model the trajectory of the shot by simply following the ball as it traveled through the air (seen below). This resulted in a quadratically shaped curve, showing the path the ball took after it made contact with my racket.

Logger pro images:

In the figure shown above, made with Logger Pro, my first impressions were that both curves made logical sense relating to my thesis. However, this is when I noticed that each picture was taken from a different angle.

'Desmos', another mathematical application, was used to identify the quadratic equation behind the curve (seen below).

$$f(x) = a(x-h)^2 + k$$

The variable (a) is a parameter which affects the shape of the parabola, vertically and horizontally stretching or contracting the parabola. In this investigation, the variable (a) will always be negative or, in other words, is concave down. This is due to the kinetic energy applied to the ball and gravity pulling it downwards. Followed by this, the variable (x) represents a value along the x-axis. This value will be manipulated further into this investigation. The variable (h) translates the parabola along the x-axis. Hence, the variable (k) translates the height of the parabola along the y-axis. The final quadratic equation is displayed in vertex form because the variables (h and k) represent the (x,y) coordinates which are the vertex or maximum point of the parabola.

D Good Reflection

When the parabola followed the trajectory of the ball as accurately as possible, the final formula was found. This equation is modeled with my table tennis shot in figure 1, depicting my forehand and figure 2 showing my backhand. The variables (a, h, k and x) were manipulated to finalize the final quadratic equation.

Reflection 1

For the collection of data, one decimal place was used because the sliders in Desmos provided them in one decimal place. Additionally, further on in the investigation when the x coordinate values are taken to find the area under the curve, one decimal place was used because it is difficult to be more accurate with the naked eye and I did not want to make any assumptions.

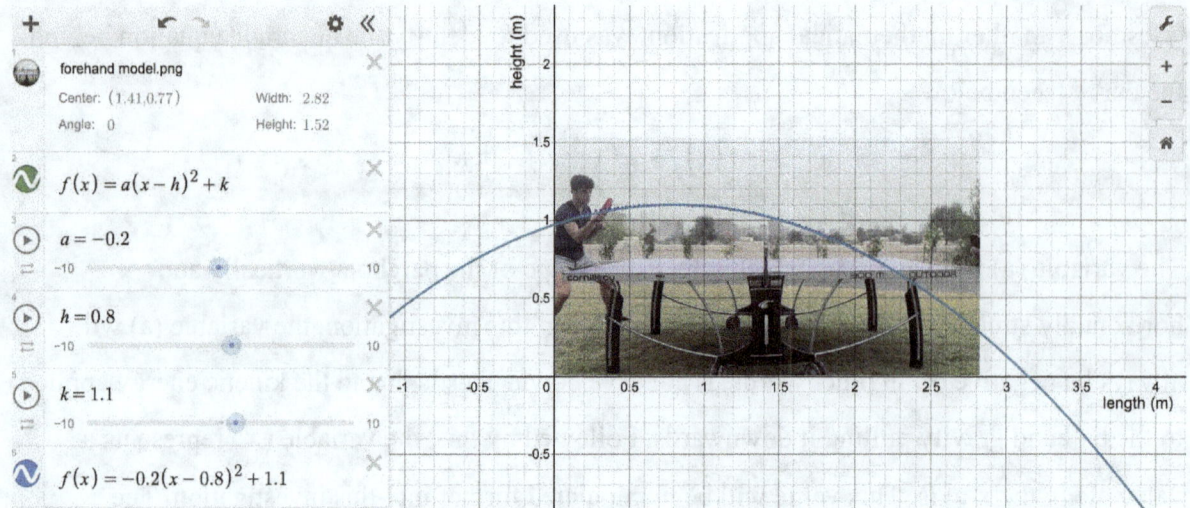

Figure 1 - Forehand shot

Figure 1 was created by uploading the image of my forehand, shown previously from 'Logger Pro' and creating a parabola which follows the path the ball took during the shot. As mentioned previously, the variables (a, h, k) were manipulated through the sliders on the left hand side of each image. Thus, figure 1 shows the path the ball takes during my forehand topspin and the quadratic equation that my short creates. The equation below models the forehand shot and figure above.

$$f(x) = -0.2(x-0.8)^2 + 1.1$$

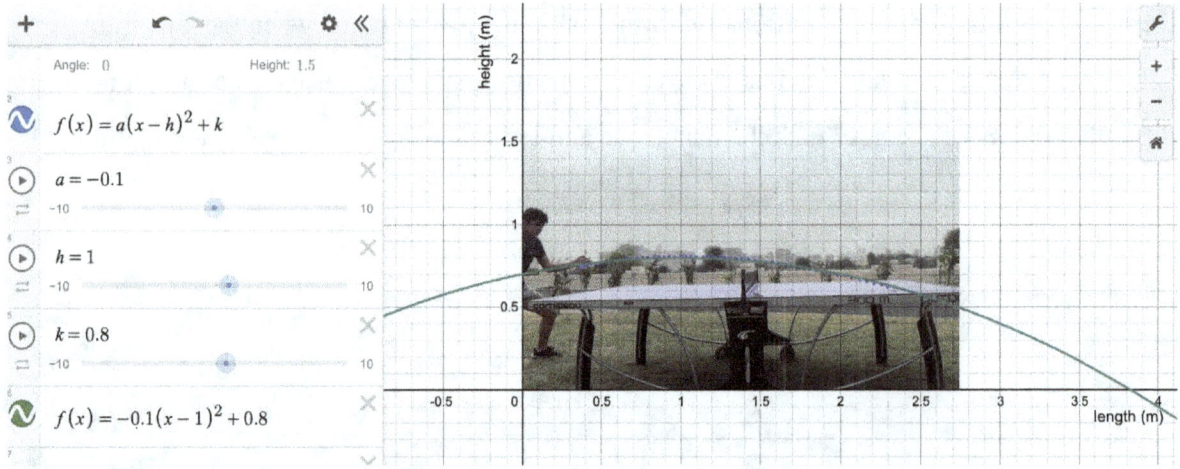

Figure 2 - Backhand shot:

Figure 2 was also made through the use of 'Logger Pro'. This image illustrates a parabola which follows the path the ball took during my backhand shot and the corresponding quadratic equation. The equation below models the backhand shot and figure above.

$$f(x) = -0.1(x-1)^2 + 0.8$$

Reflection 2:

Something to note is that one shot does not represent all my shots. My forehand and backhand shots may vary throughout different situations. Moreover, from time to time, a topspin shot will be extremely well executed and the others will not. A shot may also be less dangerous but better because it is simply more consistent. This being said, after years of training between 20-30 hours per week, I can execute both shots consistently.

More Reflection

Equations of forehand and backhand

	Forehand shot:	Backhand shot:
Equation:	$f(x) = -0.2(x-0.8)^2 + 1.1$	$f(x) = -0.1(x-1)^2 + 0.8$
Maximum point:	(0.8, 1.1)	(1, 0.8)

The maximum point was calculated by observing the (h and k) values of each equation (h) represents a translation along the x-axis and (k) represents the translation through the ($y-axis$).

Qualitative observations

From figure 1 and 2, it is prominent that the forehand shot is higher and shorter. This is due to two reasons. Firstly, when I hit my forehand, I spin the ball more than when hitting a backhand. The longer movement allows the racket to brush the ball upwards creating topspin. This projects the ball upwards, resulting in a higher (a) value and first derivative value when the ball touches the racket (will be elaborated later). Thus the high maximum point. Secondly, as the shot is hit with more topspin, the ball drops faster, resulting in a lower (h) value. The forehand shot shown in figure 1, illustrates a very short shot. High spin shots may startle low level opponents, however, experienced players know how to control the extra spin. The high arc and the short shot allows the ball to bounce high and slowly, giving the opponent more time to hit a finishing shot.

In figure 2, a backhand topspin shot is shown. Contrasting figure 1, the backhand shot has a smaller value for 'a' in its equation, making its trajectory lower and giving it a flatter curve. A backhand topspin shot usually has less spin than a forehand topspin shot as the movement is shorter and quicker. This should make it less effective than high spin shots. However, my backhand shots are taken quickly (close to the table) and at the maximum point in order to maintain safety over the net whilst hitting a quick and deep shot. Deep shots (higher (h) value) are effective as it gives the opponent very little time to hit back if they are close to the table. Logically, the lower the ball is over the net, the less time the opponent has to respond to the shot. While low shots usually have less spin, the point is won by placing the ball on the edges of the table giving the opponent little time to respond, or no chance at all if they have no space to move.

Reflection 3

Based on the above findings, the ideal shot would be low over the net, whilst having spin and deep. This equation would look similar to the backhand shot's equation, meaning that the (a) value is close to 0 (so it travels close to the net and that the (h) value would be high (so that it bounces deep on the opponent's side of the table).

Finding gradient at different points

In order to find the gradient of the curve at different points, I calculated the first derivative and substitute a value (x). By finding the first derivative of both functions, the gradient of the parabola was discovered by equating the equation to 0 and substituting (x). This is relevant and useful during this investigation as gathering the gradient of any line or curve shows you the rate of change of one variable with respect to another (in this case height and length). Calculating the rate of change between these variables at different points is relevant as it will mathematically justify if a shot is effective in its path or travels un-ideally.

Below, the quadratic parabola is broken into three pieces, point 1 (when the ball makes contact with the racket), point 2 (when the ball is over the net) and point 3 (when the ball touches the table). The (x) value inputs may be imperfect as these were taken manually by looking at figures 1 and 2. When the ball crossed each point, the video was stopped and each point was taken for both shots.

Derived equation: (forehand shot)

$$f(x) = -0.2(x - 0.8)^2 + 1.1$$

$$f'(x) = -0.4(x - 0.8)$$

Derived equation: (backhand shot)

$$f(x) = -0.1(x - 1)^2 + 0.8$$

$$f'(x) = -0.2(x - 1)$$

In order to find the stationary points of the parabola I found the first derivative of each equation before plugging in the values for (x). Once the first derivative of each equation is identified, the (x) value inputs can be substituted into the equations to find the gradient of the shot at different points;

1. When the first contact with the ball is made, when the ball touches the racket.
2. When the ball is approximately over the net and finally.
3. When the ball touches the table on the other side of the net (opponents side of the table).

1.a) Contact with racket: (forehand)

$$f'(0.3) = -0.4(0.3 - 0.8)$$

$$f'(0.3) = 0.44$$

1.b) Contact with racket: (forehand)

$$f'(0.2) = -0.2(0.2 - 1)$$

B- Presentation here could be better. MS word would help.

$$f'(0.2) = 0.16$$

2.a) Over the net: (forehand)

$$f'(1.4) = -0.4(1.4 - 0.8)$$

$$f'(1.4) = -0.24$$

2.b) Over the net: (backhand)

$$f'(1.4) = -0.2(1.4 - 1)$$

$$f'(1.4) = -0.08$$

3.a) Contact with table: (forehand)

$$f'(2.2) = -0.4(2.2 - 0.8)$$

$$f'(2.2) = -0.56$$

3.b) Contact with table: (backhand)

$$f'(2.3) = -0.2(2.3 - 1)$$

$$f'(2.3) = -0.26$$

Quantitative Observations (gradient):

The first derivative of the function represents the gradient of the parabola. A positive value signifies that at that point, the ball is moving upwards. Hence, a negative value shows that the ball has already passed its maximum point and is going downwards. Additionally, the further the value from 0 (positive or negative) the steeper the gradient at that point.

The data above suggests that during the forehand shot, when the racket hits the ball, the ball has a high gradient (0.44). The backhand shot has a much lower gradient at the point in which the racket hits the ball (0.16). A positive gradient indicates that at a point on the curve, the parabola is moving upwards. Newton's law of gravitation supports the idea that the greater the gradient is when the racket hits the ball, the higher the maximum point will be as the ball will travel at a more vertical angle. This shows that the arc of the forehand shot is consistently a lot greater than that of the backhand shot, likely due to the added topspin on the ball.

At the point where the ball travels over the net, both gradients are negative (-0.24) and (-0.08), respectively, meaning that the maximum height point of the ball is before the ball crosses the net. The forehand shot reaches its maximum point quite early due to its first derivative value being further from 0, showing that the forehand shot is slow and will likely be short. The backhand shot is a lot closer to the maximum point when it crosses the net, meaning that the shot will land deeper on the table giving the opponent less time to react.

Finally, when the ball makes contact with the table on the other side of the net (opponents side) the gradients are (-0.56 and -0.26). Suggesting that the curve is negative or traveling downwards, into the table. Although spin plays an important role in the direction in which the ball bounces, the forehand shot will bounce higher than the backhand shot due to its higher gradient. This makes the shot less effective, as a shot that bounces upwards is easier to return than one that bounces away from the table with a lower gradient.

Integration - Area under the curve:

The area between the trajectory ball and the table was calculated through the use of integration. Through the use of calculus, and more specifically through the use of integration, the area under the curve and above the table can be found. This piece of data adds validity to my investigation as the total area between the table and the net can show which shot is more likely to win the point. My assumption is that a greater area correlates to a longer duration of time before the opponent receives the ball because the ball must travel upwards against gravity, thus losing speed and giving the opponent more time to react. Thus, the area under the curve and above the table will be calculated for each shot to compare which area is greater. This will give me an indication of which shot is faster? More efficient? And makes it more difficult to return.

A limitation to this piece of data is that the area does not directly show whether a shot is better than another. Although an opponent may have more time with a shot with a large area, factors such as spin or placement could be more effective to win a point. Therefore, throughout my evaluation, I will be looking at the diagrams to aid my judgments.

In the following mathematical calculations, the (x) values are taken when the parabola touches my racket and when the parabola touches the table on the other side of the net. As seen below, for my forehand shot, the two (x) values were (0.3 and 2.2). To further understand where these two values are taken, observe these two points along the x-axis of figures (1,2,3,4 or 5).

Forehand shot:

In order to find the area under the curve for each equation, the following steps were completed to successfully integrate.

$$\int_{0.3}^{2.2} -0.2(x-0.8)^2 + 1.1 - 0.6 \, dx$$

$$\left[\int_{0.3}^{2.2} -0.2x^2 + 0.32x + 0.372 \, dx \right]$$

Firstly, the equation was expanded and then the anti-derivative or integral of each term was found. As seen below, all three terms are being integrated using the formula;

$$\frac{a^{x+1}}{x+1}$$

Don't use the word 'plug'

After each term is integrated, both (x) values (2.2 and 0.3) were plugged into the equation and subtracted in order to find the area between those two points.

$$\left[\int_{0.3}^{2.2} -0.2x^2 \, dx = 0.2 \left[\frac{x^{2+1}}{2+1} \right]_{0.3}^{2.2} \right]$$

$$\left[\int_{0.3}^{2.2} 0.32x \, dx = 0.32 \left[\frac{x^{1+1}}{1+1} \right]_{0.3}^{2.2} \right]$$

These would look better again on MS word.

$$\left[\int_{0.3}^{2.2} 0.372 \, dx = [0.372]_{0.3}^{2.2} \right]$$

Finally, all terms are added to find the total area under the curve (seen below).

$$= -0.708066 + 0.76 + 0.7068$$

$$A = 0.758734 \, m^2$$

Backhand shot

In the parabola, the ball touches my racket when (x) is (0.2) and the ball touches the table when (x) is (2.35). Same steps were taken for the backhand equation (mentioned above).

$$\int_{0.2}^{2.35} -0.1(x-1)^2 + 0.8 - 0.5 \, dx$$

$$\left[\int_{0.2}^{2.35} -0.1x^2 + 0.2x + 0.2 \, dx \right]$$

$$\left[\int_{0.2}^{2.35} -0.1x^2 \, dx + \int_{0.2}^{2.35} 0.2x \, dx + \int_{0.2}^{2.35} 0.2 \, dx \right]$$

$$-\, 0.099079 + 1.72 - 1.075$$

$$A = 0.54592 \, m^2$$

Quantitative observations (integration):

When observing the data above, one of the first things that stands out is the x coordinate of the two shots. By subtracting both x-coordinates ($x_2 - x_1$), the total distance of the shot can be measured. In the forehand shot, the ball travels approximately 190cm in distance. Whereas the backhand shot travels approximately 215cm. As stated previously, the longer a table tennis shot is, the more difficult it is for the opponent to return the ball because it gives the opponent less time to react.

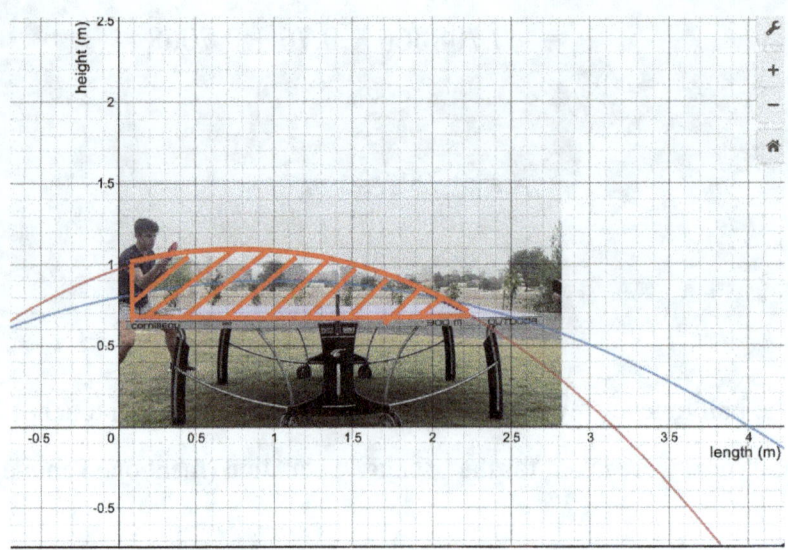

Figure 3: Area under forehand shot

In figure 3, the area found through the use of integration of the forehand topspin is striped in red. The app 'Preview' was used to create this model. To find this area the total area under the curve was found and then as seen from the equations and my working, the initial (k) value (1.1) was subtracted by (0.6) (the height of the table). This was done to calculate the area under the forehand topspin shot and above the table.

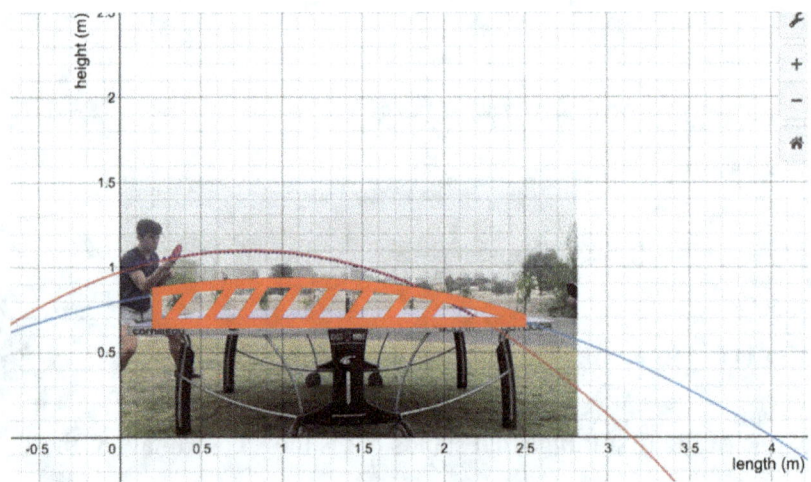

Figure 4: Area under backhand shot

C+ Student is engaging with the models

In figure 4, the area found through the use of integration of the backhand topspin is striped in red. To find this area the total area under the curve was found and then as seen from the equations and my working, the initial (k) value (0.8) was subtracted by (0.5) (the height of the table). This was done to calculate the area under the forehand topspin shot and above the table. Notice the table height in both figures is slightly different, in the forehand image the table height is about 60cm and the backhand image is approximately 50cm. I will elaborate about this limitation of data further through the investigation.

Although the area might not be very different, the forehand shot and backhand shot have very different curves. The forehand curve is shorter and higher and the backhand curve is longer and flatter. Hence the total area between the curve and the table could be similar whilst having a completely different shape. The forehand shot ($0.758734 \ m^2$) has a greater area than the backhand shot ($0.545921 \ m^2$). This is predominantly due to the high arc on the forehand topspin. Although a lower area might be beneficial to win the point, shots are usually more consistent when there is safety over the net (a bigger area).

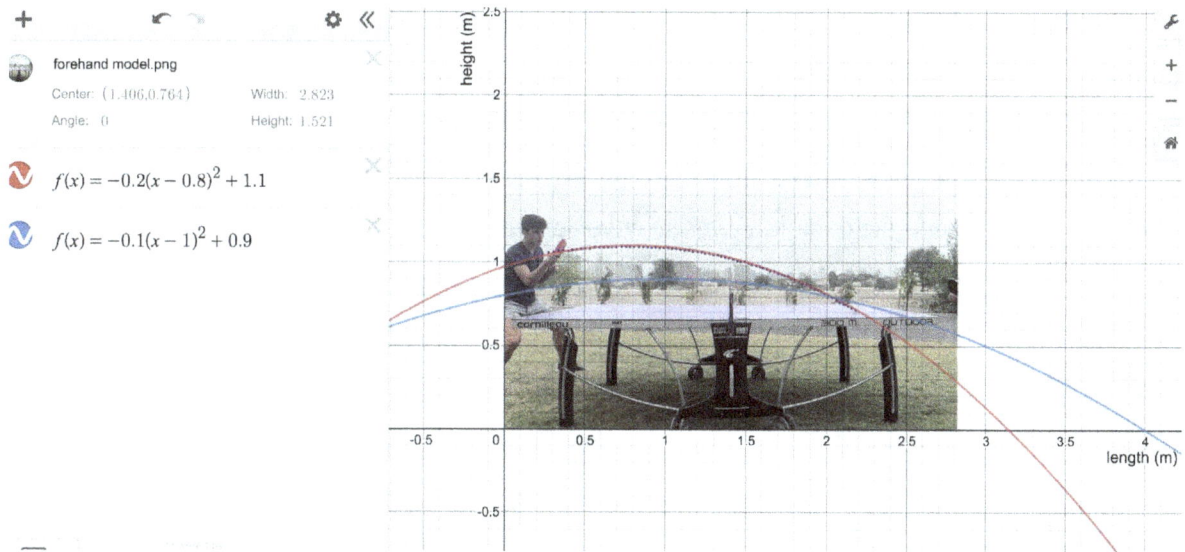

Figure 5 - Forehand and Backhand model:

Figure 5 depicts both shots (forehand and backhand) on the same model. Both equations are incorporated, as seen on the left. The blue parabola depicts the forehand shot and the green is the backhand shot. Instantly, it is evident that the forehand shot has a higher (a) value, due to the compressed shape. Moreover, the forehand shot seems to dip faster due to its shape and its higher first derivative value when the ball makes contact with the table. However, the accuracy of these judgments are limited due to the conditions the experiment was completed. The background image on figure 5 is the image taken for the forehand shot. As mentioned before, the backhand shot was taken from a different angle. Therefore, in order to fit this graph to scale, the backhand equation was altered. The (k) value was increased by (0.1) because each image was taken at a different angle. After initially merging both curves onto the same figure, the backhand curve went through the net because on the forehand model, the table was 0.6m high and on the backhand figure the table is 0.5m high. Therefore, to make both curves fit on one model, the (k) value had to be changed in order to make the diagram 'to scale'.

Initial equation:

$$f(x) = -0.1(x-1)^2 + 0.8$$

Equation in figures (3, 4 and 5):

$$f(x) = -0.1(x-1)^2 + 0.9$$

Limitations of data

I had planned to take the pictures required for this investigation in a gym (flat floor, no wind), with a better camera and a tripod (same distance and angle to the table). However, due to the Covid-19 outbreak and the containment, the environment and equipment used were not ideal in order to conduct an empirical experiment. Firstly, the shots were hit outdoors, in my garden. Wind may have altered the direction or depth of the shot at any given point. This could have resulted in an undesirable outcome as my shot would be different to the shot I hit in competitive matches. Moreover, the table tennis table was set on grass which may not be flat and my mother filmed the videos with her phone. Although I am very grateful for my mother's help, the final conclusion would have been more reliable if the angle at which each picture was taken was the same. In order to do this I could have placed my phone at net height. Due to these limitations, my backhand shot in figure 3 is going through the net. Moreover, the maximum points are fairly similar, $(0.8, 1.1)$ and $(1, 0.8)$.

D+ Good critical reflection

Another limitation to my data is that the figures and models were all made manually. For example, the path taken by the table tennis ball drawn on Logger pro was done by hand. Therefore, standard deviation, a measure of the amount of dispersion between data values, should have been used. To be more accurate, more professional equipment can be used as well as a robotically drawn graph.

Variables such as spin, speed and placement of the ball play a very important role in a table tennis shot. Therefore, by simply looking at the trajectory of a shot, we can not fully justify if one shot is more successful than another. My forehand is shorter and higher than my backhand however, there is more forward rotation on the ball (spin). One opponent might be good at returning spinny shots. However, another may be better at blocking faster shots with less spin. Therefore, a unanimous decision cannot be made because I will vary the spin, speed and depth of my shots depending on the weaknesses of my opponent.

Conclusion

This investigation intended to determine why my backhand topspin is better than my forehand topspin, in order for me to become better at table tennis. Although the angle at which the shots were taken are different, the equipment used was limited and the technology used to form models and equations was not desirable, my data holds a degree of validity. The mathematical data completed, uplifted the fact that both my forehand and backhand shots are strong.

Both shots, forehand and backhand, have their individual strengths and weaknesses. They can both be dangerous and can be vital to win a point. My forehand topspin has a higher arc due to its high rotations per second, which can baffle the opponent if they are not paying attention or are not very experienced. However, a slower shot gives the opponent more time to build up a powerful return shot. Opposingly, my backhand topspin has a lower arc, less rotations per second, but is much faster and can catch an opponent off guard. Although on paper my backhand shot may look more favorable, the high safety over the net of my forehand topspin makes it a very consistent shot. Therefore, although I might finish a point with my fast backhand, I may win more points with my forehand solely because it is more consistent.

Finally, this investigation allowed me to use quantitative data through the mathematical solving of integration, first derivatives, quadratics and graphing in order to come to a data based conclusion. The area under the curve and maximum points suggest that the backhand topspin shot is more efficient and dangerous. However, the forehand shot seems to be more consistent as it has a high margin of safety over the net, due to its high (k) value. Although I did not have the equipment to find the exact rotations per second of each shot, the forehand topspin shot seems to have more spin to hit due to the high first derivative value when it makes contact with the racket. Quantitatively, the backhand topspin shot is the better shot because it gives the opponent less time to react due to its speed and the flat path it takes in the air (low values for the variables (a) and (k)).

At the level at which I'm playing at, winning points comes down to replicating shots like this. For my future success as a table tennis player, I must train to gain consistency through my backhand shot. Whilst my backhand is the stronger shot, there is always room for improvement, as seen by the graphs located above. This can be done by practicing the shot whilst keeping the (a) value and (k) value within a similar range or further increasing the depth of the ball by increasing the value (h). Additionally, I could also improve the backhand shot by making it more spinny. For my forehand, I must work on lowering the (a) and (k) values and increasing the (h) value. This will make the forehand shot a lower and effective shot instead of a consistent shot.

Good Conclusion. Concise IA.

IA Sample 2 Examiner Comments: Dubai Ferris Wheel

A very clear, well written, concise IA where the student models the Dubai Ferris Wheel using Trig functions.

Appropriate for: AASL, AISL, AAHL, AIHL (HL would require more depth)

Criterion	SL	HL
A	4	4
B	4	4
C	3	3
D	2	2
E	6	4
Total	19	17

A: The IA is short but very well structured, with a clear aim, concise execution, and smooth flow from modelling to conclusion. The trigonometric model is introduced logically, with helpful diagrams or graphs supporting understanding. No section feels rushed or bloated; everything serves a purpose. The clarity and efficiency of the layout enhance the overall quality.

B: Mathematical expressions are clearly written, and variable definitions are consistent throughout. Functions are correctly formed and well explained. Graphs are all well presented with a title and axes labelled. Very few errors if any.

C: The topic is original, personally relevant and has a creative angle. Engagement with the mathematics is evident and steers the exploration. Student is clearly interested in the modelling, where it is going and how it will answer their aim.

D: The student reflects meaningfully on their results. There is reflection on the model's assumptions (e.g. constant speed, exact height of buildings, timing of sunset). There is plenty of reflection, but more critical reflection needed to achieve 3/3.

E: The student builds a trigonometric model of the wheel's height as a function of time. Clear and thorough understanding of this model's creation and meaning is demonstrated. They also demonstrate excellent understanding of how to solve their equation and what it means.

Introduction

The Dubai Eye is a huge Ferris wheel built on Bluewaters Isand and once completed will be the biggest in the world. I have been passing the Bluewaters Island and see the Dubai Eye almost everyday on my way to school. The constructions started on May 20th 2013, meaning it has been under construction a few days shy of six years now. I have been waiting to take a ride on it ever since they started so that I could witness Dubai with a bird's eye view and also to see the beautiful sunset. I wondered, though, if I would be in fact able to see the sunset as there are buildings in the way. You can see this in image 1 below. *Nice concise Intro. Clear aim and plan. Photo adds to it. Makes the reader interested.*

The aim of this exploration is to find the probability of viewing the sunset, while on the 'Dubai Eye'. I am defining the 'sunset' as the moment the sun touches the horizon. My plan is to find out all the information I can get about the Dubai Eye and model the height at which I would be when riding the Ferris wheel with respect to time. If I can find out how long I will be above the buildings, I should be able to calculate the probability of seeing the sunset.

Image 1: Dubai Eye seen on the left of the Image. Photo taken by me.

Creating the model

To model the height of the Dubai eye with respect to time, I was aware that I could apply a trigonometric function. As the Ferris wheel rotates, the height of a capsule (with me in it), goes up and down. As the Ferris wheel ride begins the height would increase which lead me to think of the sine function as it increases from 0.

A+ Very clear. Writing is simple. Helps the reader understand.

Using Desmos, an online graphing calculator, I sketched $h = \sin t$ (figure 1 below) to get an initial idea of what it might look like. For all graphs, t (time) will be the horizontal axis and h (height) will be the vertical axis.

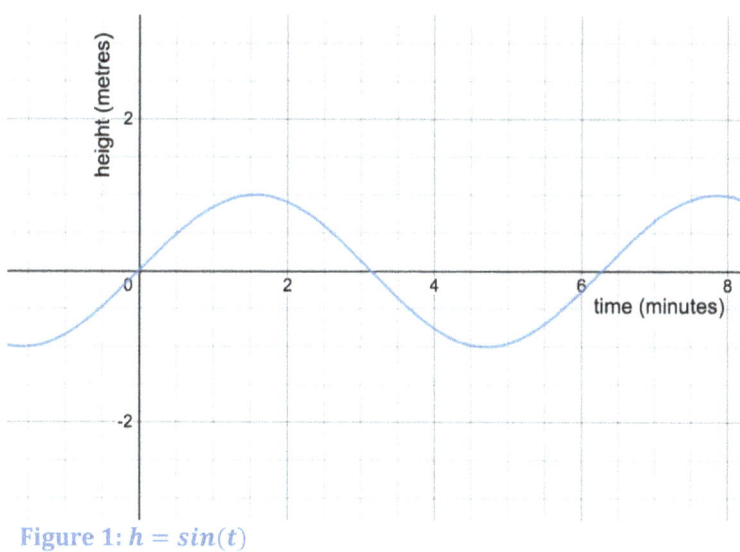

Figure 1: $h = sin(t)$

Immediately after looking at the graph, I realized I needed to know some information about the ferris wheel. To calculate the amplitude, I needed to the length of the diameter. To calculate the period, I needed to know how long it takes for the wheel to do a complete turn.

B+ D+ Clear image, axes labeled then student immediately reflects

To find out this information, I searched online. Unfortunately there is no official website yet for the Dubai Eye. The best source I could find was an Emirates247 news article. However, even though this is a respectable Dubai newspaper, the information it gives may not be perfectly accurate as 1) it is not an official source and 2) the Dubai Eye is still under construction so there is no way to be certain how long a full rotation would take. The article told me that the diameter of the wheel would be 250m. It said the time for one full rotation would be 45 minutes. It also provided me with another important piece of information that I didn't think about. The ride will begin 12m off the ground.

To begin transforming my sine function into a model of the Dubai Eye, I decided to first calculate the amplitude. By dividing the diameter of the wheel by two I can get the correct amplitude, which is 125m. I then graphed $h = 125\sin(t)$. This is shown below in Figure 2.

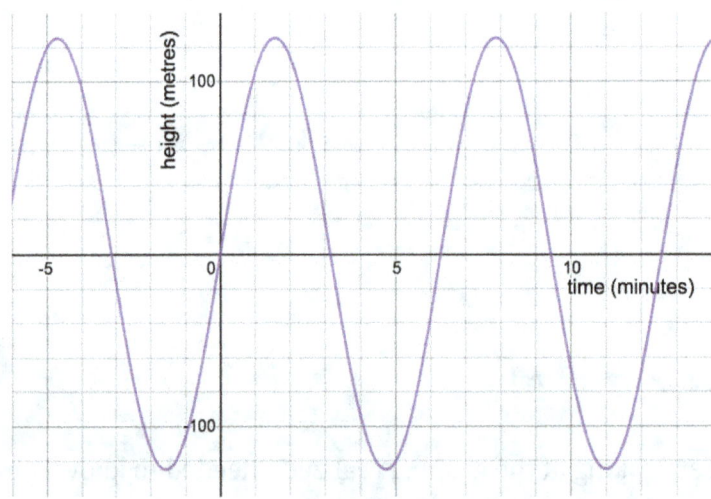

Figure 2: $h = 125\sin(t)$

More good reflection and writing is again very easy to follow

At this point I could see I was starting to get somewhere. However according to this graph, the Ferris wheel goes underground. This is obviously not the case. I needed to vertically translate it up 125m as, at the moment, the minimum is at -125 and I need it to be at 0. I then also remembered that it will start 12m above the ground. I instead chose to translate it up 137. 125 + 12 for the starting height. To do this I added 137 to the function, giving $h = 125\sin(t) + 137$.

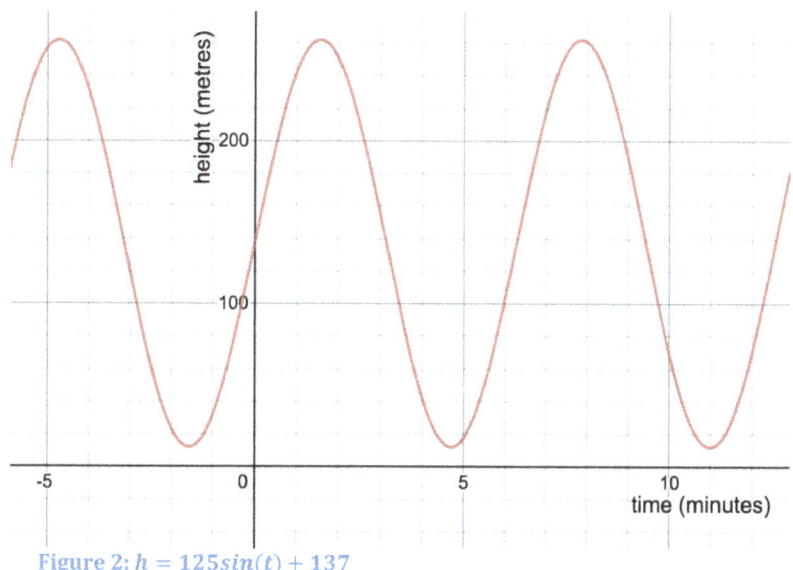

Figure 2: $h = 125\sin(t) + 137$

This is now a more realistic representation of the Dubai Eye but the period needs to be 45 minutes because this is time needed for a full rotation. To achieve a 45-minute period on the graph, I applied a horizontal stretch. Using the formula $\frac{2\pi}{b} = period$ for a sine function of the form $h = \sin(bt)$, I can change the period to 45.

$$\frac{2\pi}{b} = 45$$

$$b = \frac{2\pi}{}$$

This will give the function $h = 125\sin\left(\frac{2\pi}{45}t\right) + 137$, shown in figure 4.

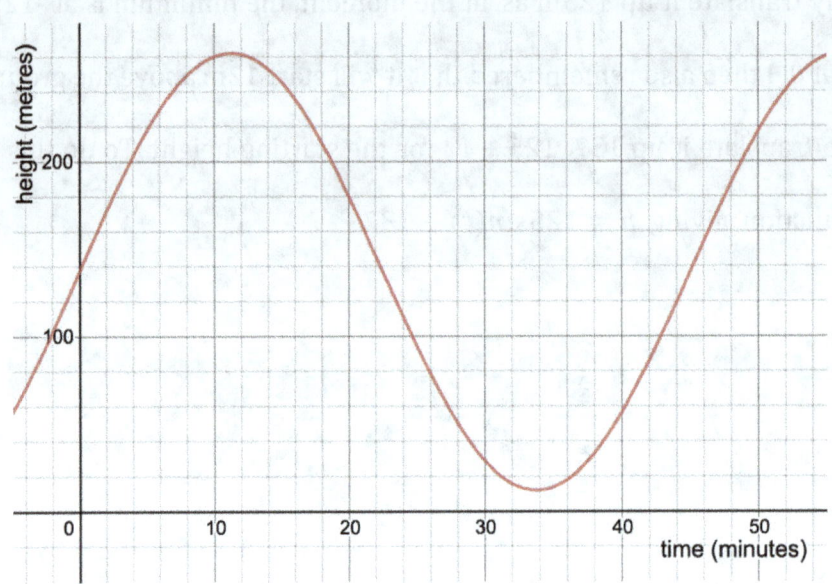

Figure 3: $h = 125\sin\left(\frac{2\pi}{45}t\right) + 137$

This is very close to being accurate but at $t = 0$, the time at which the Ferris wheel ride begins, the height should start from 12 meters and not 137 meters.

In order to fix this, I need to translate the graph to the right. But by how much? The maximum height occurs at 11.25 minutes as this is a quarter of the period. This means if I translate the graph right by 11.25, the minimum should occur at 0. This will give the function:

This is an example of the maths, results and reflection guiding the exploration. After each graph, the student lets the reader know what they think and where they now want to go.

$$h = 125\sin\left(\frac{2\pi}{45}(t - 11.25)\right) + 137$$

> E+ Entire modelling process was very well and simply explained. Student demonstrated a very good understanding of transformations of graphs.

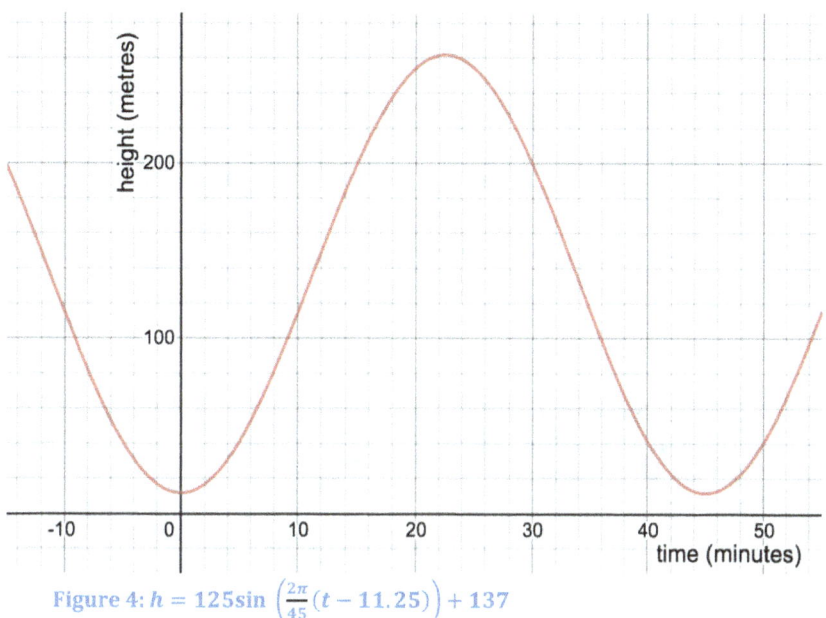

Figure 4: $h = 125\sin\left(\frac{2\pi}{45}(t - 11.25)\right) + 137$

This looks great and is the model I wanted.

Reflection.

I genuinely enjoyed this process. I have always found transformations very difficult and applying them to a real life situation with the help of Desmos really helped my understanding of the topic. While I had been introduced to Desmos before, I never really knew how to use it properly. Through this exploration, I feel I have learned a new, very useful and valuable skill. While I do think this is a pretty good model, it is not perfect as the data we have used is not perfectly accurate.

> More reflection although this paragraph is a little superficial.

Comparing height with buildings.

Now I have my model, I can find the time when I can see the sunset. The reason I can't see the sunset at every height on the Dubai Eye is that there are buildings in the way. This is shown in image 2 below. It is a photo of a model of the Bluewaters Island, which I took. Image 3 is a screenshot from Google maps. It shows Bluewaters

Image 2: Model of Bluewaters Island. Taken by me. Image 3: Google maps view of Bluewaters Island

Island with the Dubai Eye. You can see that the buildings are to the west, which is where the sun will set.

To solve the problem, I need to know the height of the buildings. To simplify the problem, I decided to find the height of the tallest building and use this as the minimum height I would need to be, to see over the buildings. I have a friend who lives on the island and she told me that the height of the tallest building on the island is 63 meters. I believe my friend but it should be noted that she is not a completely reliable source.

Example of how photos can add to your IA. Helps the reader stay interested.

If I equate my function to 63 and solve the equation, I can find the times at which I will be above the tallest building and hence be able to see the sunset.

My initial equation is given by,

$$125 \sin\left(\frac{2\pi}{45}(t - 11.25)\right) + 137 = 63$$

Subtracting 63 gives,

$$125 \sin\left(\frac{2\pi}{45}(t - 11.25)\right) = -74$$

And dividing by 125,

$$\sin(\frac{2\pi}{45}(t - 11.25)) = \frac{-74}{125}$$

$$\sin\left(\frac{2\pi}{45}(t - 11.25)\right) = -0.592$$

Using inverse sine on my calculator,

B+ Example of really good Mathematical communication. Equations in the centre, explanations to the right, plenty space between.

$$\frac{2\pi}{45}(t - 11.25) = -0.633538$$

And solve for t,

$$-4.53738 = t - 11.25$$

$$t = -4.53738 + 11.25$$

$$t = 6.7126$$

$t = 6.712$ is the first time the height will be above 63m. To find the time the height returns below 63m, I need to subtract 6.712 from 45 as the graph is symmetrical. I will define the time at which the height first goes above the building as t_1 and the time at which it returns below as t_2.

$$t_2 = 45 - 6.712 = 38.288$$

These times are given in minutes and are rounded to 5 significant figures. As we are dealing with time, I am interested in rounding only to the nearest second. To convert the decimal to seconds, I just need to multiply the decimal by 60.

$$0.7126 \times 60 = 42.972$$
$$0.288 \times 60 = 17.28$$

Therefore rounded to the nearest second,

$$t_1 = 6 \; minutes \; 43 \; seconds$$

and,

$$t_2 = 38\ minutes\ 17\ seconds$$

These results mean that after I start my ride on the Dubai Eye, I will be below the tallest building and therefore unable to see the sunset for the first 6 minutes and 43 seconds. After this time, I will be above the tallest building until 38 minutes and 17 seconds have passed, after which time I have returned below again.

Reflection

I found solving this trigonometric equation to be the hardest part of the exploration. But again, I think the process and relating it to my own example has improved my understanding of the topic. A big limitation with my solution is that I cannot be certain that the tallest building is 63m exactly. To improve the accuracy, I could have tried to get in contact with the developers to ask for more accurate values.

Finding the probability

After all this work, we still haven't found the probability of seeing the sunset which is our aim. To do this, all I need is to find the total time I will be above the buildings and divide that by 45 as this is the time for one complete turn. To find the time spent above the buildings, I just subtract t_1 from t_2.

$$t_2 - t_1 = 31.576$$

This means that the probability of seeing the sunset on the Dubai Eye is

$$\frac{31.576}{45} = 0.7016888\ldots = 0.702 \ (3 \ significant \ figures)$$

This is the answer to our aim. I now know the probability of me seeing the sunset. 0.702 is given to 3 significant figures as this gives enough accuracy, especially considering the limitations of my model. In fact, seeing as it is so close to 0.7, I could even round to 1 decimal place. If someone asks me what is the probability of seeing the sunset on the Dubai Eye, I would probably say 'around 70%'.

However, this gives the probability of seeing the sunset if I go on the Dubai Eye at any random time. If I really want to make sure I see the sunset, I could increase this probability by planning what time I start the ride. In fact, again assuming my model is correct, if I know the time of the sunset, I just need to make sure I start the ride at least 6 minutes and 43 seconds (t_1) before sunset and no more than 38 minutes and 17 seconds (t_2).

Page numbers missing but very simple, concise, well written IA

Conclusion

The aim of this exploration was to find the probability of seeing the sunset on the Dubai eye. In order to calculate this, I created a mathematical model to find out how long I would be above the tallest building and hence able to see the sunset. The probability I calculated was 70%. This means if I jump on the Dubai Eye at a random time and stay on until the sun sets, I am more likely than not, to see the exact moment that the sun sets. I also figured out that I could use the information I calculated to plan when to get on the Ferris wheel so I would definitely see it. Perhaps this would have been a better aim.

There were many limitations with the accuracy of my result. For starters, as the Ferris wheel is not even built, there was no way to get perfectly accurate data to build my model. Also, I am assuming a constant 45-minute period or time for full rotation. In reality, Ferris wheels take different amounts of time depending how busy it is and if everything is working perfectly. I also assumed all buildings were as tall as the highest one which is obviously not true. I could have improved the exploration by looking at exactly which buildings were in the way at which point on the Dubai Eye. Perhaps when it is open, I can actually time how long the ride takes and how long I can see the horizon and see if I get a similar result.

While my work may not be perfectly accurate, I thoroughly enjoyed working on this and it has really helped my understanding of trigonometric modelling and mathematical modelling in general.

IA Sample 3 Examiner Comments: Volume of an Egg

An excellent HL IA that models and finds the volume of the Cadbury's Conundrum egg.

Appropriate for: AAHL, AIHL

Possible for AASL, AISL (Maths could be more basic for SL)

Criterion	SL	HL
A	4	4
B	4	4
C	3	3
D	3	3
E	6	5
Total	20	19

A: The IA is clearly structured, and very well written. It flows well and is very easy to follow for a peer. Graphs and diagrams are well presented. It is concise.

B: All variables are clearly defined, units are stated and consistent, and functions are labelled. Equations are presented using proper notation and look good. Different models are distinguished clearly, and each is given space to be properly explained before being compared. Graphs are neat, labelled, and interpreted.

C: The IA is very original. The student demonstrates ownership in multiple ways. They compare different models. They independently research the University of Kent egg formula. They interpret the result in a creative and original way. The motivation and curiosity are evident throughout.

D: There is regular reflection throughout. The student consistently reflects on model fit and limitations. They discuss which models approximate the egg well and why. They show a good understanding of assumptions (e.g. assuming rotational symmetry, ignoring surface features). They critically assess the mathematical outcome (e.g. how different models affect the volume and thus the value).

E: The IA uses a wide range of HL mathematics including volume of revolution and modelling with polynomials and ellipses. The egg formula adds a nice bit of sophistication, and it is explained well. Explanations of mathematics are excellent and demonstrate a very good overall understanding. All techniques are applied correctly and interpreted meaningfully.

Title: Modeling the Cadbury Conundrum Egg to calculate its material value

Could have had a cooler title

Subject: Mathematics Analysis and Approaches HL

No need for subject

Page count: 15

Introduction

In 1983, the chocolate manufacturer Cadbury created 12 eggs made of 22-carat gold and hid them around Britain for a treasure hunt. A thirteenth, larger egg – the Conundrum Egg - was presented to one lucky Cadbury retailer in the UK. In February 2021, the family of the retailor auctioned the egg, which was sold for £372,000 (Russel, 2021). This paper will investigate the optimum model for this egg in order to calculate the material value of the gold used to create the egg. I chose to investigate this topic after reading an article on the auction of the Conundrum Egg. While I understood that the egg was a limited-edition art piece which had gained value over time by being so rare, I was nonetheless surprised by the price it fetched. I decided to figure out how much if this price tag could be attributed to the actual price of the gold, and how much was due to the collector's value.

The aim of this investigation is to find the most appropriate model for the Conundrum egg, and then use this model to calculate the volume of the egg. Using the volume, it is then possible to find the value of the specific amount of gold in the egg, and then find the total price of the egg's material cost. I will consider 3 models of the Conundrum Egg: a simple elliptical model, a more intricate polynomial model, and finally, a specialized model using the universal "egg formula", developed by researchers at the University of Kent, which is slightly more accurate than the previous models.

Model 1 – Ellipse

A+ Concise intro with clear aim and plan

For my first model, I considered the egg in its most simple possible form– an ellipsoid. In order to plot a model for the egg, I taught myself how to use GeoGebra, an online graphing software. Using this allowed me to insert a picture of the Conundrum Egg onto the axes, which I made to-scale (8.3cm tall, 6.35cm wide at its widest point), as seen in Image 1. This will then allow for a model to be plotted onto the egg on GeoGebra, which can then be used to find the volume of the egg using the formula for a volume or revolution about the x-axis.

Image 1: Conundrum Egg on GeoGebra, alterated to scale – Egg picture from (Batemans Auctioneers and Valuers, 2021)

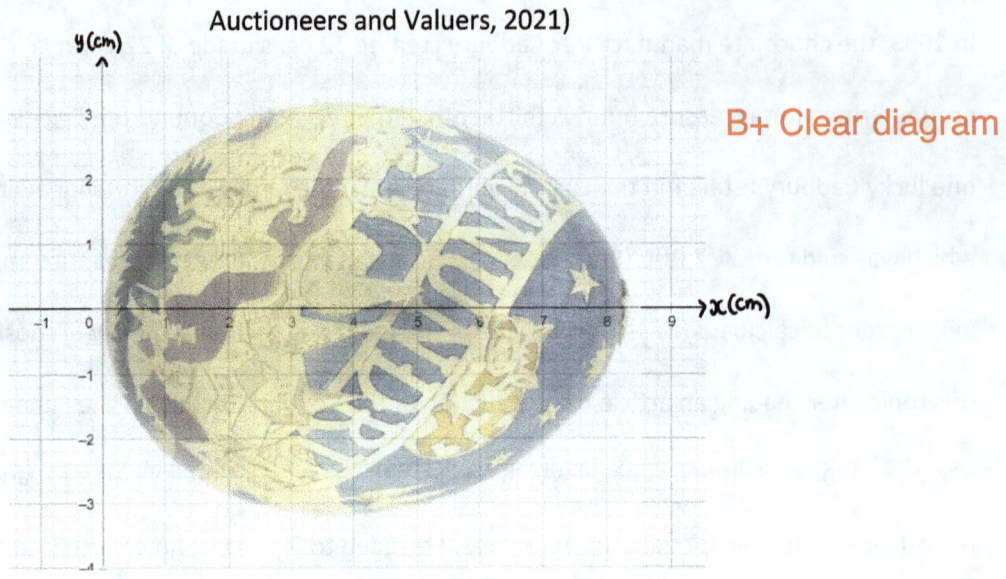

B+ Clear diagram

The standard equation for an ellipse is:

$$\frac{(x-h)^2}{a^2} + \frac{(y-k)^2}{b^2} = 1$$

Where the center of the ellipse is (h, k), the horizontal length of the ellipse is $2a$ and the vertical length of the ellipse is $2b$. Image 2 shows these values annotated on the image of an ellipse.

Image 2: Ellipse annotated with standard equation values

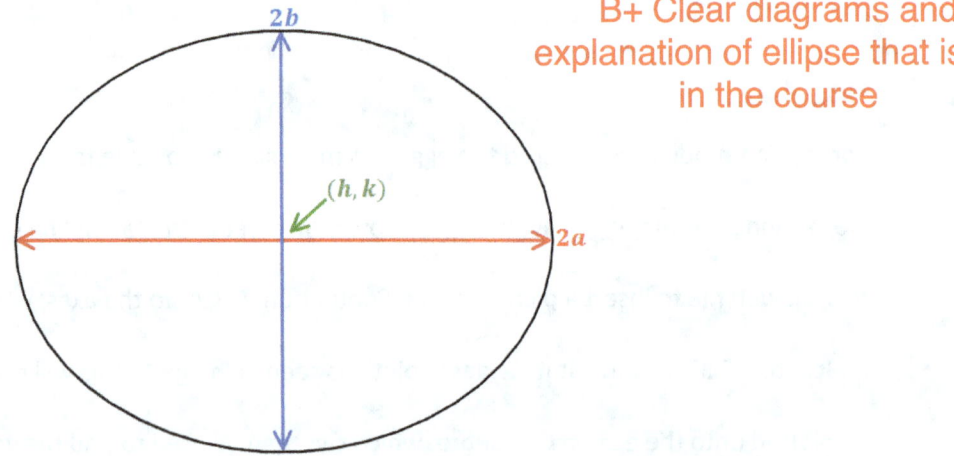

B+ Clear diagrams and explanation of ellipse that is r in the course

The center coordinates (h, k) of the egg on GeoGebra were $(4.15, 0)$, as can be seen in Image 1.

Knowing that the maximum width of the egg is 6.35cm and its length is 8.3cm, the values are set as:

$$2a = 8.3$$

$$2b = 6.35$$

$$h = \frac{8.3}{2} = 4.15$$

$$k = 0$$

Substituting these values into the standard equation:

$$\frac{(x-4.15)^2}{\left(\frac{8.3}{2}\right)^2} + \frac{y^2}{\left(\frac{6.35}{2}\right)^2} = 1$$

$$\frac{(x-4.15)^2}{17.2225} + \frac{y^2}{10.0806} = 1$$

Image 3: Ellipse equation plotted onto image of Conundrum Egg on GeoGebra

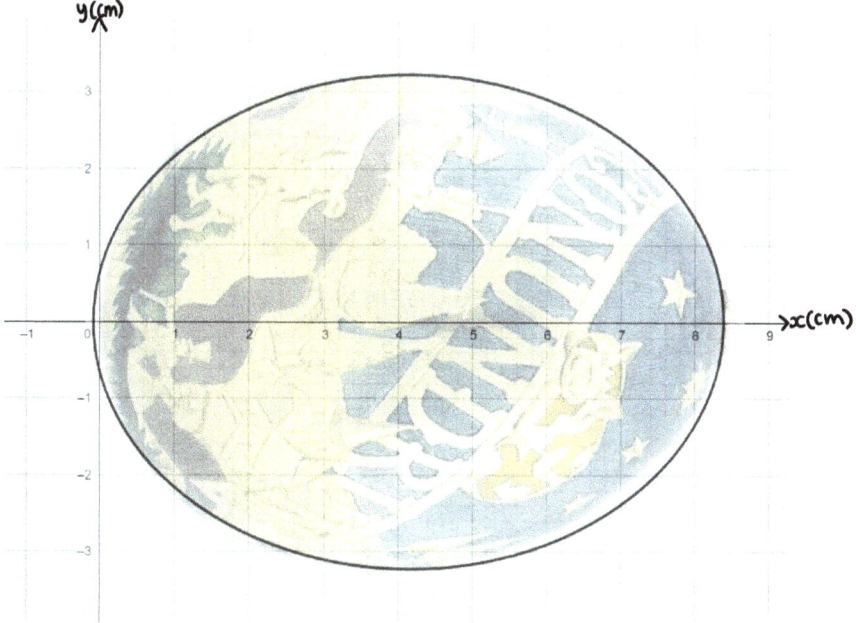

After inputting this onto GeoGebra, I altered the values slightly to get the best possible ellipse fit for the egg, ending up with a final equation of:

$$\frac{(x-4.15)^2}{17.94} + \frac{y^2}{10.4} = 1$$

However, the formula of an ellipse is a relation but not a function, as it has more than one y output for each x input, and it does not pass the vertical line test. To integrate the model and find its volume, I rearranged for y to make it a function and defined the equation as $f(x)$:

$$\frac{y^2}{10.4} = 1 - \frac{(x-4.15)^2}{17.94}$$

$$y^2 = 10.4\left(1 - \frac{(x-4.15)^2}{17.94}\right)$$

$$y = \sqrt{10.4\left(1 - \frac{(x-4.15)^2}{17.94}\right)}$$

$$f(x) = \sqrt{10.4\left(1 - \frac{(x-4.15)^2}{17.94}\right)}, \quad 0 \leq x \leq 8.3$$

B+ E+ This section is very well explained and very well presented

Image 4 below shews this elliptical model plotted onto the Conundrum Egg on GeoGebra. While this not a perfect model, it provides a good basis for a rough estimate of the shape.

Image 4: Elliptical model of the Conundrum Egg on GeoGebra

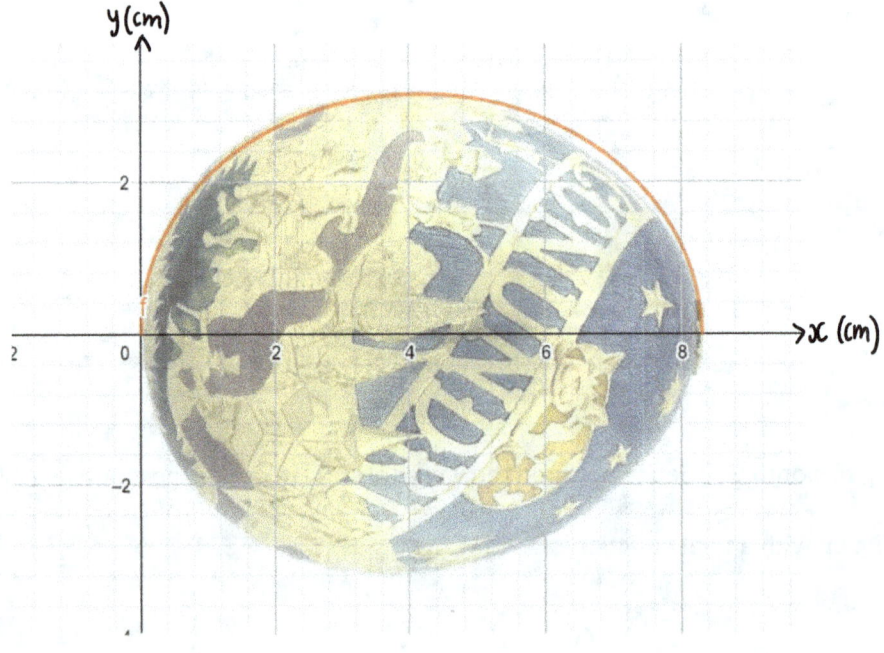

The general formula for volume of a revolution about the x axis is:

$$V = \int_a^b \pi y^2 \, dx$$

Inputting the equation for $f(x)$:

$$V = \pi \int_0^{8.3} (f(x))^2 \, dx$$

$$V = \pi \int_0^{8.3} 10.4\left(1 - \frac{(x-4.15)^2}{17.94}\right) dx$$

Integrating analytically:

$$V = 10.4\pi \left[x - \frac{(x-4.15)^3}{3(17.94)}\right]_0^{8.3}$$

$$V = 10.4\pi \left(\left(8.3 - \frac{(8.3-4.15)^3}{3(17.94)}\right) - \left(0 - \frac{(0-4.15)^3}{3(17.94)}\right)\right)$$

$$V = 10.4\pi(5.6439845782238) \approx 184.403 cm^3 \ (6\ s.f.)$$

Solving with a GDC to verify my answer:

$$\pi \int_0^{8.3} 10.4\left(1 - \frac{(x-4.15)^2}{17.94}\right) dx \approx 184.403 cm^3 \ (6\ s.f.)$$

Reflections on Model 1

While I am not able to quantify how well the model reflects the egg shape, is a pretty good fit. Yet, it is obvious that the elliptical model is not a perfect egg shape – the egg is vertically asymmetrical (wider on the bottom than the top), while the ellipse is symmetrical. As seen in Image 5, the model had quite a bit of overlap with the egg at some points and left some gaps at others. While these errors were not

Image 5: Zoom-in of Model 1

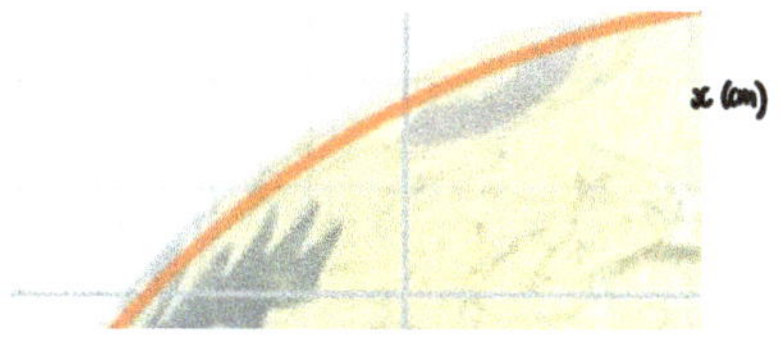

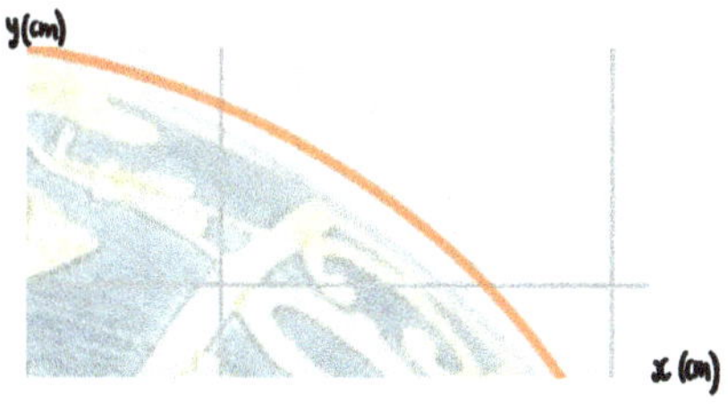

D+ Good reflection including zoomed in diagram to reflect critically

significant, they still affected the accuracy and reliability of the elliptical model, so I knew I had to find a different model.

A+ This is a good example of a nice tranistion (flow) from one section to the next.

Model 2 – Polynomial

The ellipse was by no means a perfect model, so I wanted to find a model that could replicate the asymmetrical shape of the egg. This is why, for my second model I chose a polynomial equation. After plotting a set of points (C to F_1 in Image 6) around the circumference of the egg using the Point Tool, I used the Polynomial Fit tool on GeoGebra to find a suitable polynomial equation to model the egg around these points. I tried 4th, 5th and 6th degree polynomials, but I found that the 6th degree was the best fit model, defined as $h(x)$ below. The 4th degree and 5th degree polynomials left significant gaps between the egg and the line that would have drastically affected the volume of the model, making them inaccurate. The formula for $h(x)$ shown below is rounded to 5 decimal points, as this was the smallest number of decimal points the function could be rounded to while still keeping its shape.

$$h(x) = -0.00080x^6 + 0.02018x^5 - 0.19815x^4 + 0.96271x^3 - 2.51995x^2 + 3.76065x + 0.24648, \quad 0 \leq x \leq 8.3$$

The polynomial model $h(x)$ is shown below in Image 6.

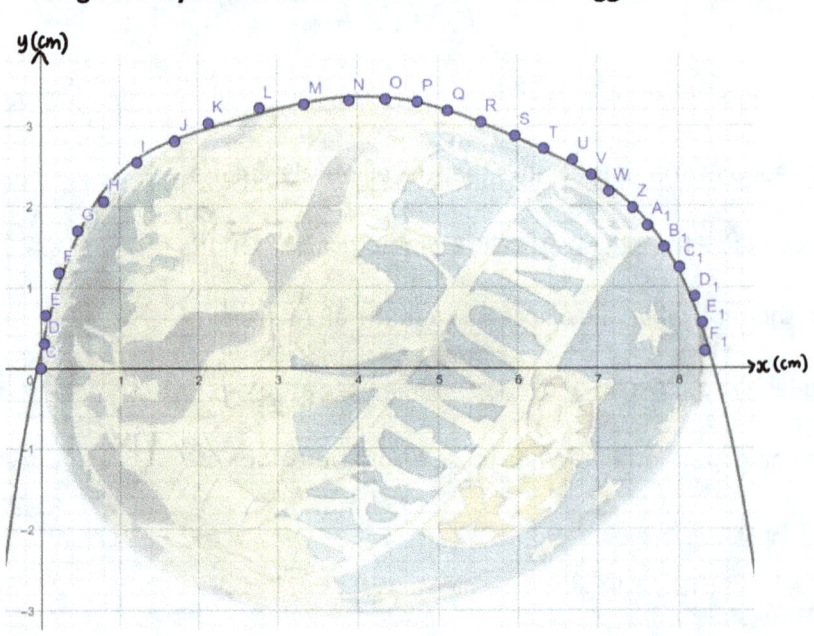

Image 6: Polynomial model of the Conundrum Egg on GeoGebra

Substituting this into the formula for volume of a revolution about the x-axis:

$$V = \pi \int_0^{8.3} (h(x))^2 \, dx$$

Solving with a GDC: E+ A+ Appropriate to solve this with the GDC

$$V \approx 191.743 cm^3 \ (6 \ s.f.)$$

Reflections on Model 2

To check the model's fit, I calculated its R^2 value, which is a statistical measure of how well the model passes through the data on the graph. The R^2 value was 0.97 (2 $s.f.$), indicating an adequate fit, though one must note that the data points were plotted manually and are not 100% accurate themselves. However,

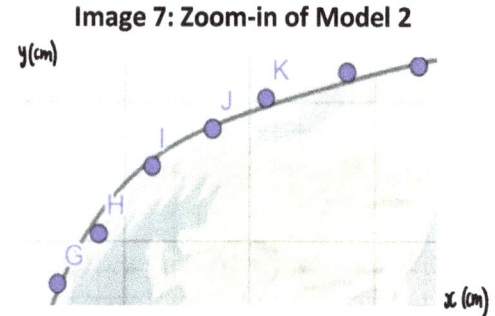

Image 7: Zoom-in of Model 2

Model 2 is a not necessarily better fit for the egg than Model 1. While are less gaps between the line and the egg and it follows the egg's curve more uniformly, this is still not a very suitable model. The line $h(x)$ does not curve in the same way as the egg and is quite jagged and unsmooth, leaving a few gaps and overlaps between the line and the egg, as shown in Image 7. Evidently, the volume of this model would be just as inaccurate, if not more inaccurate, that that of the first model, and it is not as near to exact volume as I would have liked to get. Thus, I went on to find a third, better model.

A+ D+ More good reflection and flow between sections

Model 3 – Egg Equation

After creating the last 2 models that were not great fits for the egg, I decided to research more into the shape, curvature, and symmetry of an egg. During my research, I came across a general "egg formula" from the University of Kent (Narushin, Romanov, & Griffin, 2021). This is an ideal equation which graphically models the shape of an egg based on three parameters: its length L, its maximum breadth B, and its contour w. Contour is the outline of a curved or irregular figure (Merriam-Webster). The egg formula is:

$$y = \pm \frac{B}{2}\sqrt{\frac{L^2 - 4x^2}{L^2 + 8wx + 4w^2}}$$

The contour w is defined as:

$$w = \frac{L - B}{2n}$$

Where n is a parameter affecting the egg's shape ($n > 1$). The greater the value of n is, the smaller w is, and the more elliptical the shape becomes. If $L = B$, then $w = 0$, and the shape is a perfect ellipse. Image 8 below shows 3 different contours of an egg depending on the value of n. By overlaying the Conundrum egg onto figures A, B and C (Image 9), I concluded that Model A best fits the egg's shape, and reflects the curvature almost perfectly. Therefore, I let $n = 2$.

Image 8: Contour of eggs depending on the value of n (Narushin, Romanov, & Griffin, 2021)

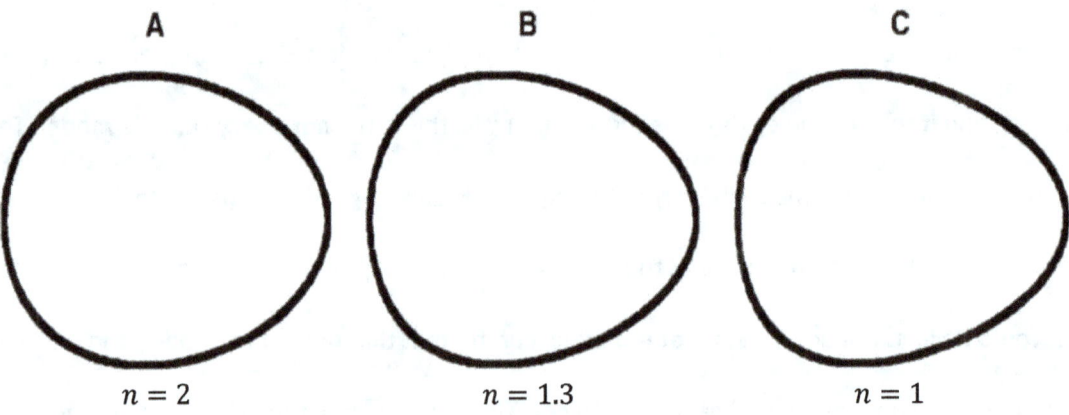

Image 9: Contour of eggs depending on the value of n with the Conundrum Egg overlayed

A+ B+ Diagrams really help explain this new concept

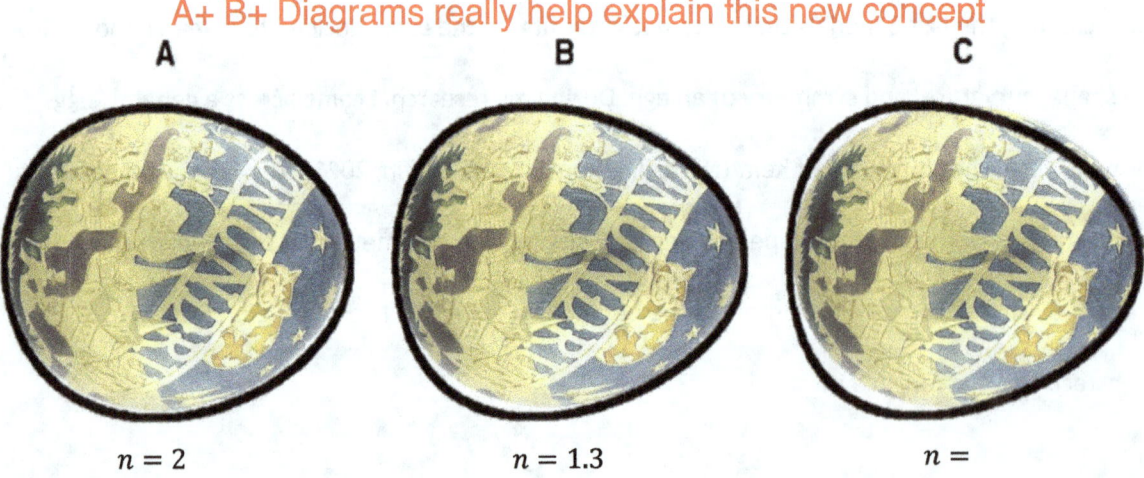

For the Conundrum Egg, the values are defined as:

$$L = 8.3 \text{ (length of egg)}$$

$$B = 6.35 \text{ (maximum breadth of egg)}$$

$$n = 2 \text{ (taken from the curvature models in Images 8 \& 9)}$$

Image 10 below shoes these values plotted onto the Conundrum Egg.

Image 10: Values of L and B shown on the Conundrum Egg

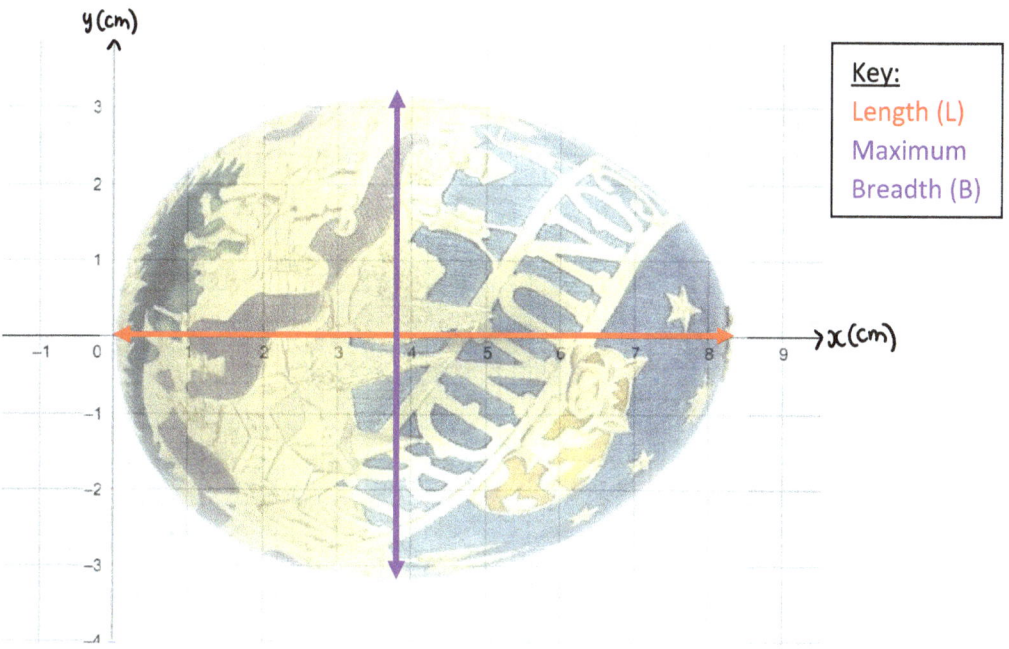

Substituting these n, L and B values into the equation:

$$w = \frac{8.3 - 6.35}{2(2)} = 0.4875$$

Setting $y = g(x)$ and substituting in w, L and B values:

$$g(x) = \pm \frac{6.35}{2} \sqrt{\frac{8.3^2 - 4x^2}{8.3^2 + 8(0.4875)x + 4(0.4875)^2}}$$

However, this formula assumes the center of the egg to be (0,0), so it must be horizontally shifted by 4.15 units to the right to make the center of the egg (4.15,0). To do this, the equation of the Conundrum Egg will have to be $g(x - 4.15)$, signifying a horizontal shift of $g(x)$ to the right by 4.15 units, therefore the equation becomes:

$$g(x - 4.15) = \pm \frac{6.35}{2} \sqrt{\frac{8.3^2 - 4(x - 4.15)^2}{8.3^2 + 8(0.4875)(x - 4.15) + 4(0.4875)^2}}$$

However, the $\pm \frac{6.35}{2}$ at the start of the equation means this is two y outputs for every x input. Similar to the ellipse model, this is a relation rather than a function, as shown in Image 11 below.

Image 11: "Egg Formula" plotted on the Conundrum Egg on GeoGebra

In order to plot this as a function that can be integrated, I set the start of the equation to $+\frac{6.35}{2}$, which then only plotted the upper part of the curve, as shown in Image 10. The equation will therefore become:

$$g(x - 4.15) = +\frac{6.35}{2} \sqrt{\frac{8.3^2 - 4(x - 4.15)^2}{8.3^2 + 8(0.4875)(x - 4.15) + 4(0.4875)^2}}, \quad 0 \leq x \leq 8.3$$

Image 12 shows the model $g(x - 4.15)$ plotted onto the Conundrum Egg on GeoGebra.

Image 12: "Egg Formula" function of the Conundrum Egg on GeoGebra

Thus, calculating the volume of the egg using the egg formula:

$$V = \pi \int_0^{8.3} (g(x))^2 \, dx$$

Solving with a GDC:

$$V \approx 174.752 cm^3 \; (6 \, s.f.)$$

This value for the volume of Model 3 is around $10 cm^3$ less than the volume of Model 1, and around $17 cm^3$ less than Model 2. This means that, considering Model 3 to be the most accurate of the three, the first two models over-estimated the volume of the Conundrum Egg, and Model 2 was actually comparatively worse than Model 1. This is most likely due to the gaps between the lines of Model 1 and Model 2, as reflected upon through Images 5 & 7, which would have led to a calculated volume greater than the actual volume of the egg.

I then used the 3D modeling tool on GeoGebra to create a 3D rendering of Model 3, as shown in Image 13 below.

Image 13: 3D model of Model 3 on GeoGebra

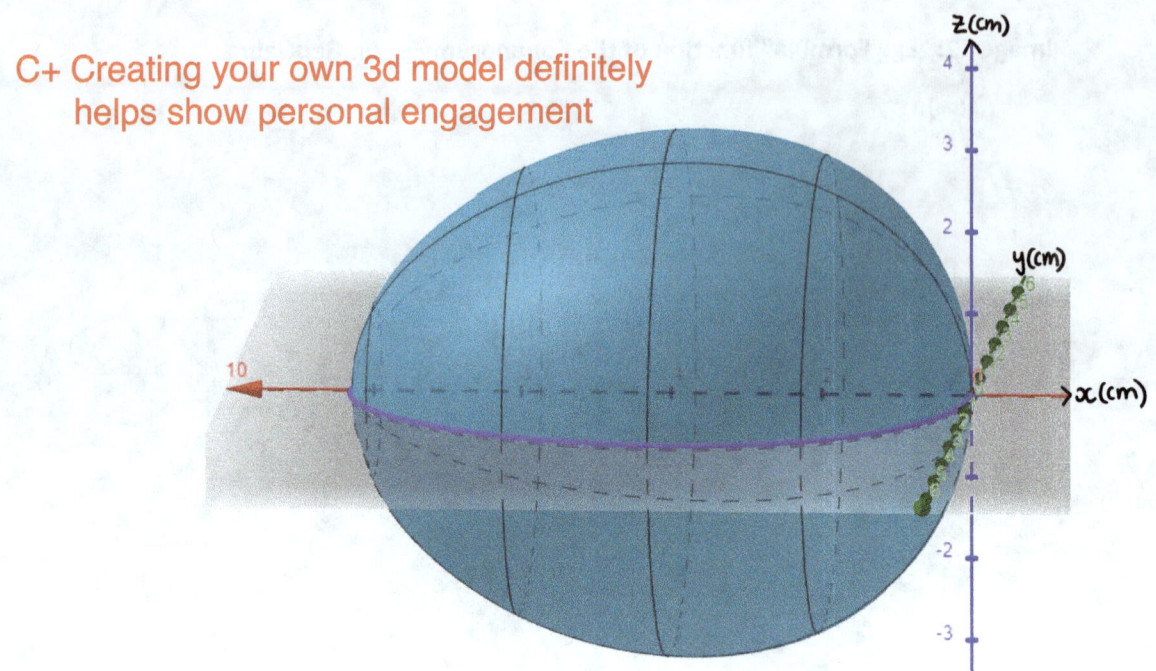

C+ Creating your own 3d model definitely helps show personal engagement

Reflections on Model 3

Comparing this model to the first and second models in this investigation, Model 3 is the best fit for the Conundrum Egg – it follows the curvature of the egg almost perfectly, and leaves barely any gaps that could affect the calculated volume of the egg. The function ideally models the egg's wider base, narrower top, and its curvature. However, the Egg Formula model of the Conundrum Egg is still not 100% accurate; there are still a few very small gaps, as pictured in Image 14 above. Yet, the extremely small size of these gaps means that the extent of these errors could be deemed as minimal, if not negligible, as they are unlikely to have a noticeable impact in the calculations for the egg's volume.

Image 14: Zoom-in of Model 3

D+ More good reflection

Material Value Calculations

In order to calculate the value of the gold, I used the volume of the egg found in Model 3 ($174.752 cm^3$) as it was the most accurate figure for the volume that was found of the 3 models. Then, using this value for the volume of gold, I calculated the mass of the gold using the density of gold ρ, defined as:

$$\rho = \frac{m}{V}$$

Where m is the mass of gold in grams g, and V is the volume of cold in cm^3. The density of 22 carat gold is $15.6 g/cm^3$, therefore, solving for mass:

$$\frac{m}{V} = \frac{m}{174.752} = 15.6$$

$$m = (174.752)(15.6) = 2726.13 g \ (6 \ s.f.)$$

Therefore, my estimate for the mass of the Conundrum Egg is approximately $2726.13g$.

However, the price of gold is quite volatile. Therefore, to calculate the material value of the egg, I found the price of gold at the time of its auction to achieve the most accurate value. The price of 22 carat gold per gram in the UK in February of 2021 (the date of the egg's auction) was around £41.33 (Bullion Rates). Using this value, I calculated that the price of the gold in the Conundrum Egg as:

$$Price = (Price \ per \ gram)(Mass \ in \ grams) = (41.33)(2726.13) = £112671 \ (6 \ s.f.)$$

My estimate for the value of (in 2021) the gold used to make the Cadbury Conundrum Egg came out to be £112671.

Conclusion and Evaluation

In conclusion, the results showed that the material value of the gold used in the Conundrum Egg was £112671, a little under one third of the price it fetched at its auction. Evidently, this is just an estimate for the value of the gold using the models created in this investigation; it is likely that the percentage of gold is slightly more or slightly less, therefore making this value of the egg not entirely representative, but merely an approximated value.

This investigation had numerous strengths; for example, the utilization of GeoGebra, an online graphing platform, allowed me to create to-scale models and manipulate the equations until they were the best they could possibly be. Moreover, the use of the "Egg Formula" further improved the modelling in this investigation to produce a very accurate and representative final model. Finally, the trials of three different models led me to finding the most suited model rather than using the simplest or most straight-forward model of the Conundrum Egg.

However, there were also a few limitations throughout this investigation. One limitation is that the image of the Conundrum Egg that I used for the models was taken from the internet, meaning that it could have been taken at an angle that slightly deformed the egg's shape or made it appear different, or could have been edited, thus causing the models to be less accurate. Secondly, even the final model, Model 3, had a few slight errors, as seen in Image 11 and previously reflected on. While these errors may have had only a minimal effect on the calculations of the volume and then the material value of the egg, it is important to note that these slight errors were in fact present.

This has to be one of the best IAs I have read

Bibliography

Batemans Auctioneers and Valuers. (2021). Large 22ct gold Cadbury's 'Conundrum' egg. doi:https://www.batemans.com/catalogue/lot/B6973EFF2F803BA7AD2569B81DD3B4B0/9F72C1814B6F84555A7FE5BC5D559A1B/jewellery-watches-silver-gold-lot-137/

Bullion Rates. (n.d.). Gold Price History in British Pounds (GBP) for February 2021. Retrieved November 1, 2022, from https://www.bullion-rates.com/gold/GBP/2021-2-history.htm

Merriam-Webster. (n.d.). *Contour.* Retrieved November 4, 2022, from Merriam Webster Dictionary: https://www.merriam-webster.com/dictionary/contour

Narushin, V., Romanov, M., & Griffin, D. (2021, August 23). Egg and math: introducing a universal formula for egg shape. *Annals of the New York Academy of Sciences, 1505*(1), 169-177. Retrieved October 24, 2022, from https://doi.org/10.1111/nyas.14680

Russel, S. (2021, February 24). Golden Cadbury's egg sold at auction. *Yahoo! Finance.* Retrieved October 1, 2022, from https://ca.finance.yahoo.com/news/golden-cadbury-egg-sold-auction-150001495.html

IA Sample 4 Examiner Comments: Turtle House Area

This student ended up buying a new house for their turtle because of this IA. It is an excellent exploration where the student uses integration to calculate the area of their turtle's house.

Appropriate for: AASL, AISL, AAHL, AIHL

Criterion	SL	HL
A	4	4
B	4	4
C	3	3
D	2	2
E	5	4
Total	18	17

A: The IA is very clearly structured, beginning with a clear personal aim, followed by model construction, area calculation, and conclusion. It flows very well and is an easy read for a peer. The layout presents well and makes it easy for the reader to follow each stage of the investigation. Everything is relevant to the aim making it concise.

B: Functions are written clearly with appropriate notation. Integration steps are well explained and clearly set out. Graphs and diagrams are all clear, titled and labelled and look good. Units are used properly when interpreting area.

C: The IA is clearly rooted in the student's personal context and motivation — they photographed their own turtle's enclosure and used this to construct a model. The student takes ownership of the IA. The student is clearly engaged with the maths and lets it guide the exploration. Unique methods of modelling the enclosure help with Personal Engagement. The fact that the student ends up buying a new turtle house really helps with Personal Engagement.

D: Reflection is present around the accuracy and usefulness of the model. There was critical reflection, and the student discusses the limitations of the model. 3/3 was possible but 2/3 was conservatively awarded. If the student could quantify their confidence in the results, this would have helped.

E: The IA uses multiple functions to model a shape and applies definite integration to find total area. It is done correctly, and the student demonstrates good understanding. The student combines these ideas meaningfully and accurately. The mathematics is appropriate and relevant to the aim. Again, 6/6 was possible but the student maybe needed to demonstrate a bit more understanding of the modelling.

Math AA SL

Internal Assessment

Good title.

Are my Turtles Happy?

An investigation of the area of my turtle's enclosure using the area under the curve

Page count: 16

Candidate code:

Introduction

As the owner of two pet turtles, it is important to me that they have a comfortable and suitable living environment. I want to ensure that their enclosure provides the necessary space for them to thrive, both in terms of physical movement and overall well-being. I was interested in finding out how much space they have in their current turtle house. I have a uniquely shaped turtle house, which you will see shortly. It is curved in a peculiar way. I thought about using my newly acquired knowledge of integration to find the area. This would help me decide if my turtles have enough space and whether I should get them a new home. **This exploration aims to calculate the area of my turtles' enclosure to determine if it is large enough to house two turtles comfortably.**

I planned to photograph the turtle house and place it on a set of axes to scale. Using functions, I would model the outline of the house. Then, using integration, I can calculate the total area. The only criterion I will use to determine whether the enclosure is big enough is the opinion of the vet. Once I calculate the area, I will ask my vet if this is sufficient space and then conclude whether the turtle house is sufficiently big.

A+ Short, well written Intro with clear aim, rationale and plan

Modeling the Turtle House

The first step was to put a photo of the turtle house onto a set of axes to scale. I used the online graphing calculator, Geogebra. I took an aerial view photo of the enclosure and imported it into Geogebra. I placed it on a set of axes to scale by ensuring the distances on the graph equal the real-life lengths of the turtle house. I did this by measuring the length of the turtle house. I found it was 30cm. As I only used a normal ruler, my measurements were not perfectly accurate, although I am confident they were correct to the nearest cm as I measured it 4 times, and each time it was between 29.8cm and 30.2cm. Figure 1 shows the turtle house on a set of axes to scale. You can see that the house starts at 0 and ends at 30.

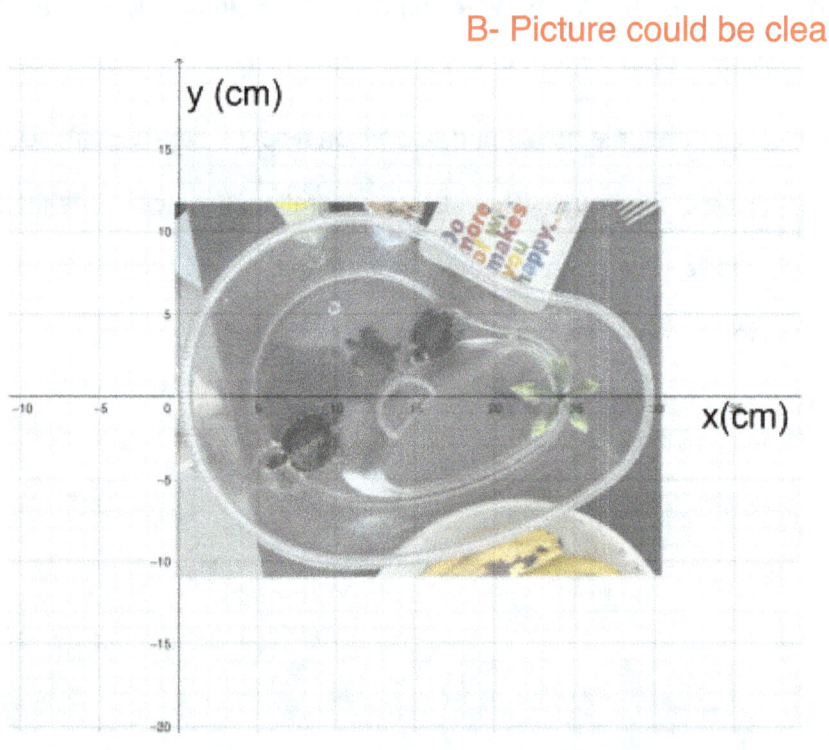

B- Picture could be clearer

Figure 1 - Turtle house on a set of axes to scale

I decided to place the center of the turtle house along the x-axis, this is because it is not symmetrical, and I would need to model both sides separately to calculate the area accurately. I spent time thinking about what kind of function would best model the upper half of the house. My first thought was a polynomial, as the graphs of polynomials have turning points. I needed a graph (which you can see from the outline of the turtle house in Figure 1) that increases to a maximum and then decreases. However, the concavity needs to change from concave down to concave up to concave down. My first thought was a negative polynomial of degree 4, as this is how its graph generally behaves. To find a function that fits, I placed points along the outline of the house. I was then able to use Geogebra's computer power to come up with a function that best fit the points and, hence, the outline of the house. Geogebra uses polynomial regression to find a curve that best fits the points. The points can be seen in Figure 2.

A+ E+ Explanations are clear. I know what they are trying to do.

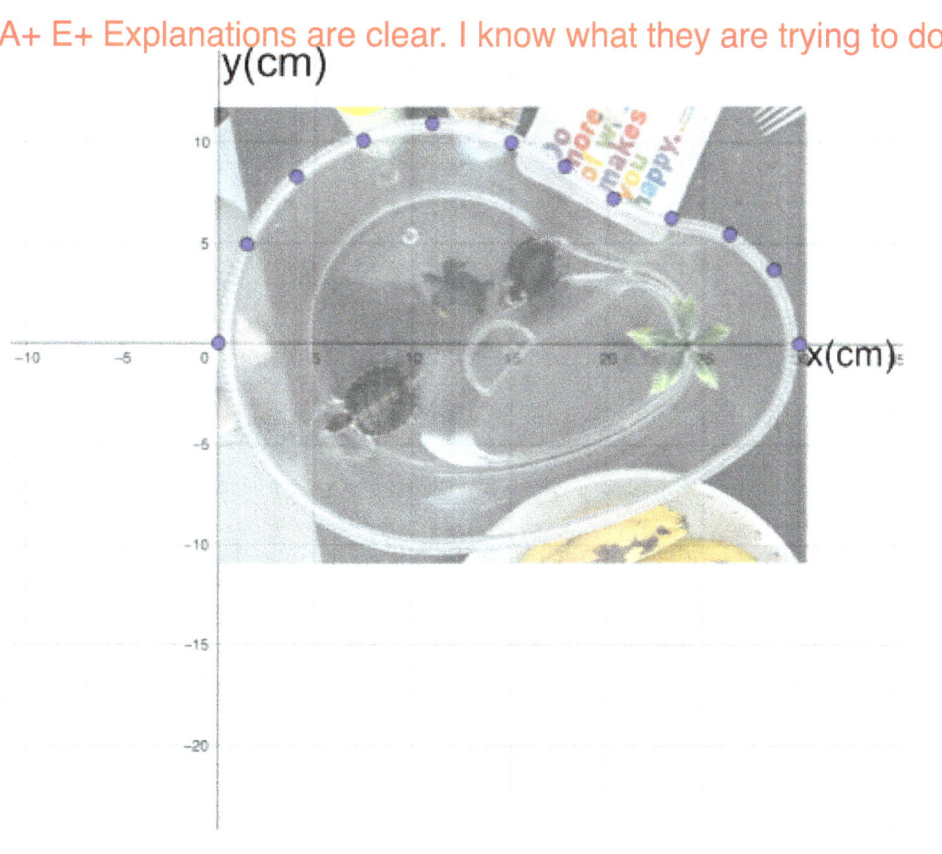

Figure 2 - Screenshot from Geogebra showing the points for the equation

Below in Figure 3, you can see the polynomial of degree 4.

Figure 3 - Polynomial of degree 4

The graph in Figure 3 is the graph of

$$y = -0.00017897x^4 + 0.012230x^3 - 0.30813x^2 + 3.0831x + 0.32918$$

You can see from the graph that while my idea to use a polynomial function of degree 4 was reasonable, the graph does not fit along the edge of the house quite how I was hoping for. I decided to increase to a degree 6 polynomial as this would provide the possibility of a better fit. I skipped degree 5 as I needed an even degree polynomial. This is because I needed a polynomial that increased at the beginning and decreased at the end. Odd-degree polynomials do not do this. The graph of the degree 6 polynomial can be seen below in Figure 4.

D+ E+ Good reflections on the models and explanations about why the functions were chosen

Figure 4 - Polynomial of degree 6

The graph in Figure 4 is the graph of

$$y = -0.0000019345x^6 + 0.00016907x^5 - 0.0056672x^4 + 0.093105x^3$$

$$- 0.83211x^2 + 4.2445x + 0.060090$$

Note that I set the coefficients to 5 significant figures. When I set them to less, the function did not match what Geogebra initially gave. 5 significant figures matched almost exactly.

This is a much better fit than the polynomial of degree 4, and it was reasonable to now try to calculate the area. However, I still felt I could do better before integrating. I felt I could find a graph or graphs that could fit around the outline better. You can see at the beginning and end of the turtle house (i.e., when $x = 0$ and $x = 30$) that the graph doesn't quite fit onto the outline. I decided to split the outline into two different functions. This would allow me to try and model using different types of graphs and not be limited to just a single polynomial. I recognized that the first part could be modeled using a circle. I researched how to find the equation of a circle and, by changing the center and radius, came up with the circle shown in Figure 5.

More good reflection

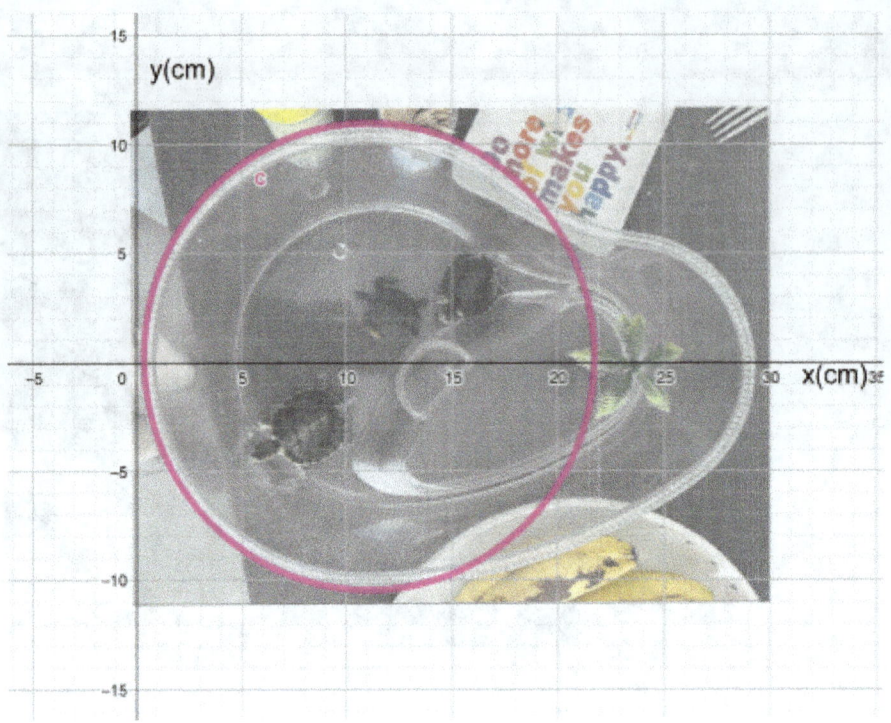

Figure 5 - Circle

The general equation of a circle is given by the equation.

$$(x - a)^2 + (y - b)^2 = r^2$$

Where (a, b) is the centre of the circle and r is the radius. The equation of a circle can be derived from the distance formula where the distance from the centre of the circle (a, b) to a general point on the circumference (x, y) is given by:

$$\text{Distance} = \sqrt{(x - a)^2 + (y - b)^2}$$

As the distance from the centre to the circumference is the radius, we can substitute distance for r. Squaring both sides gives us.

$$r^2 = (x - a)^2 + (y - b)^2$$

B+ E+ Mathematical communication and understanding are both well demonstrated here

The equation of the circle in Figure 5 is

$$(x - 11.100)^2 + (y - 0.22190)^2 = 115.44$$

This fits the first part well, although I needed to rearrange the equation to make sure it was in the form of a function.

$$(y - 0.22190)^2 = 115.44 - (x - 11.100)^2$$

$$y - 0.22190 = \sqrt{115.44 - (x - 11.100)^2}$$

$$y = 0.22190 + \sqrt{115.44 - (x - 11.100)^2}$$

Note that I used the positive part of the root as I needed the upper half of the circle. After limiting the domain to $0.35765 \leq x \leq 18.238$, I got the below graph. I chose not to start at 0 as the house started slightly to the right of the origin.

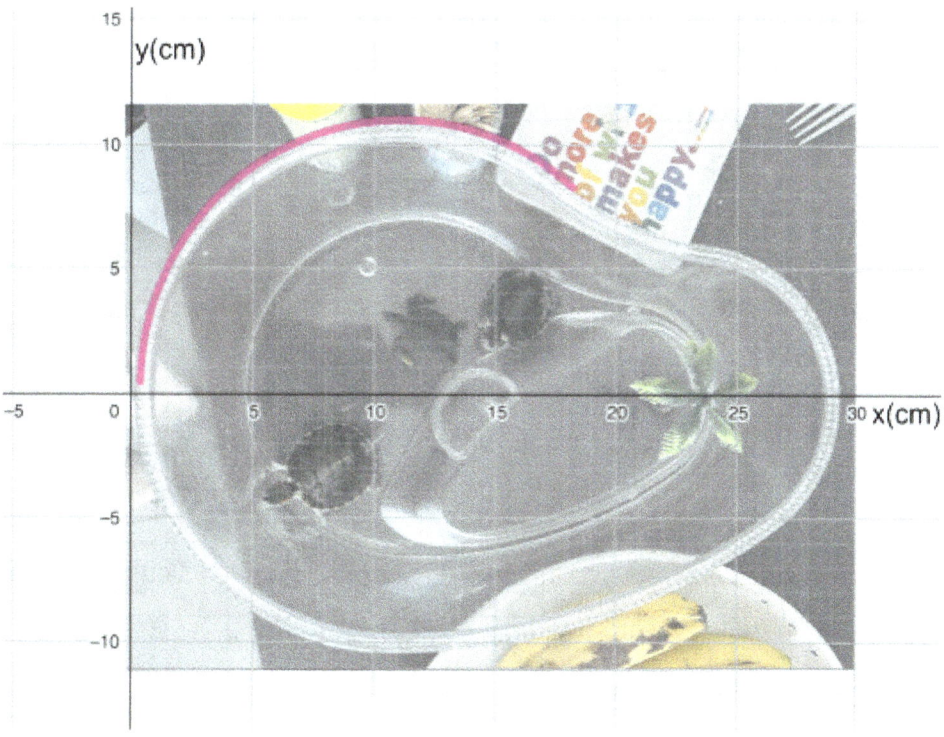

Figure 6 - Circle with its limitations to the border of the enclosure

This was a good fit. To model the rest of the upper half, using a similar method to previously, I used a polynomial of degree 3. Geogebra gave me the following function

$$y = -0.010838x^3 + 0.73993x^2 - 17.007x + 138.06$$

I also limited the domain of this function to $18.238 \leq x \leq 30$. This gave the graphs in Figure 7 below.

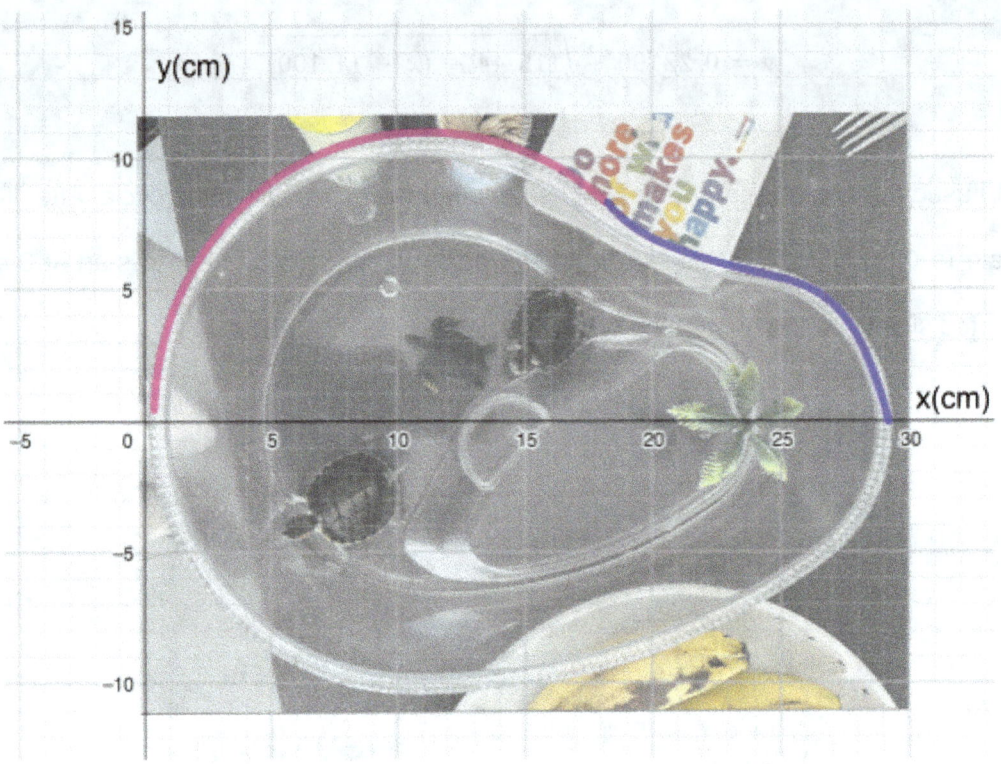

Figure 7 - Polynomial and circle with limited domains

I was happy with this model. Next, to find the area, I integrated both functions using the domain limits as the integral limits, using the formula for the area under a curve.

$$A = \int_a^b f(x)dx$$

The area between the pink graph and the x − axis is given by

$$\text{Area 1} = \int_{0.35765}^{18.238} \left(0.22190 + \sqrt{115.44 - (x - 11.100)^2}\right)dx$$

$$\text{Area 1} = 165.24 \, cm^2$$

For this integration, I used my calculator and double-checked with a separate online calculator.

The area between the blue graph and the $x-$axis is given by

$$\text{Area 2} = \int_{18.238}^{30} \left(-0.010838x^3 + 0.73993x^2 - 17.007x + 138.06\right)dx$$

After integrating,

$$= \left[-0.0027095x^4 + 0.24664x^3 - 8.5035x^2 + 138.06x\right]_{18.238}^{30}$$

Substituting the limits,

$$= \left(-0.0027095(30)^4 + 0.24664(30)^3 - 8.5035(30)^2 + 138.06(30)\right) -$$

$$\left(-0.0027095(18.238)^4 + 0.24664(18.238)^3 - 8.5035(18.238)^2 + 138.06(18.238)\right)$$

Which gives,

$$\text{Area 2} = 67.40 \, cm^2$$

As this was a polynomial, I knew how to integrate analytically, and therefore, I have shown the steps.

To find the total area of the upper half of the turtle house, I added these two areas together.

$$\text{Upper Area} = \text{Area 1} + \text{Area 2}$$

$$\text{Upper Area} = 165.24 + 67.40$$

$$= 232.64 \, cm^2$$

The next step was to model the other half of the turtle house. I decided to rotate the image so that my functions would be positive or above the $x-axis$. This would keep calculations simpler as the integration will give positive values. The image below shows this rotation.

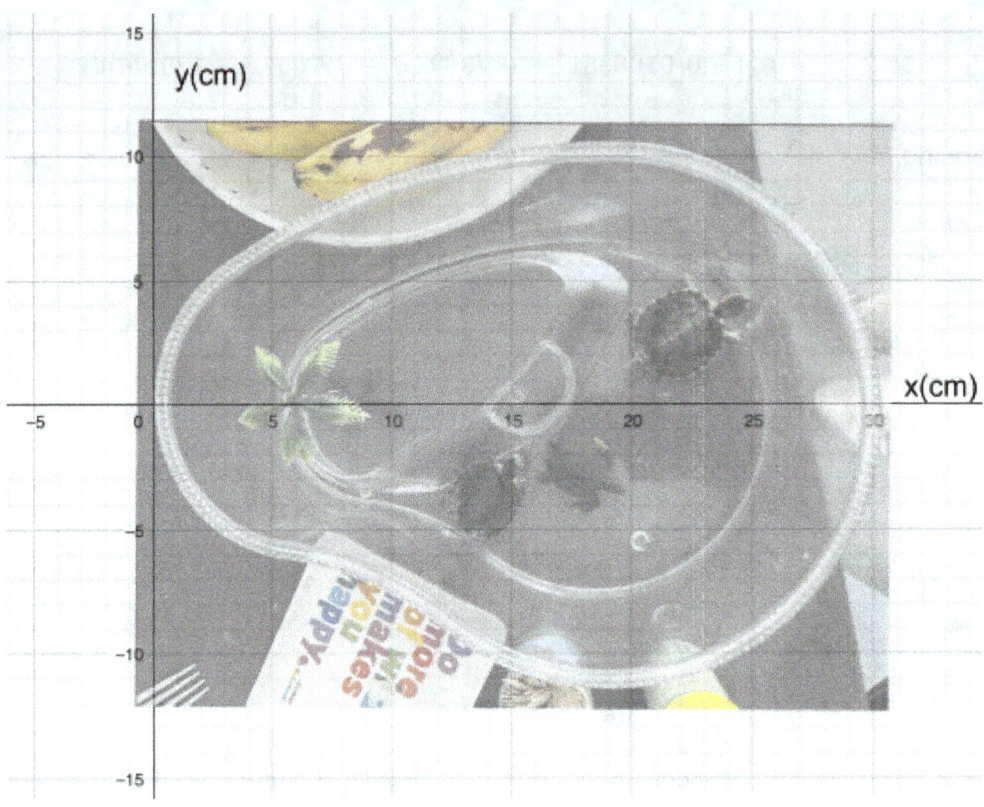

Figure 8 - The bottom half of the enclosure to scale

Again, I needed to decide which functions to use. I decided to break it up into three different parts. The gradient of the first part of the turtle house (from 0) is very high but quickly decreases. I considered again using a circle, but then I remembered the log function does this. It has a high positive gradient around 0, and then the gradient quickly decreases. Again, using Geogebra's regression tool, I found the following graph.

> A+ C+ D+ E+ Here you can see how all the criteria are linked.
> Good coherent explanations and reflections of the mathematics
> show personal engagement engagement and guide the exploration

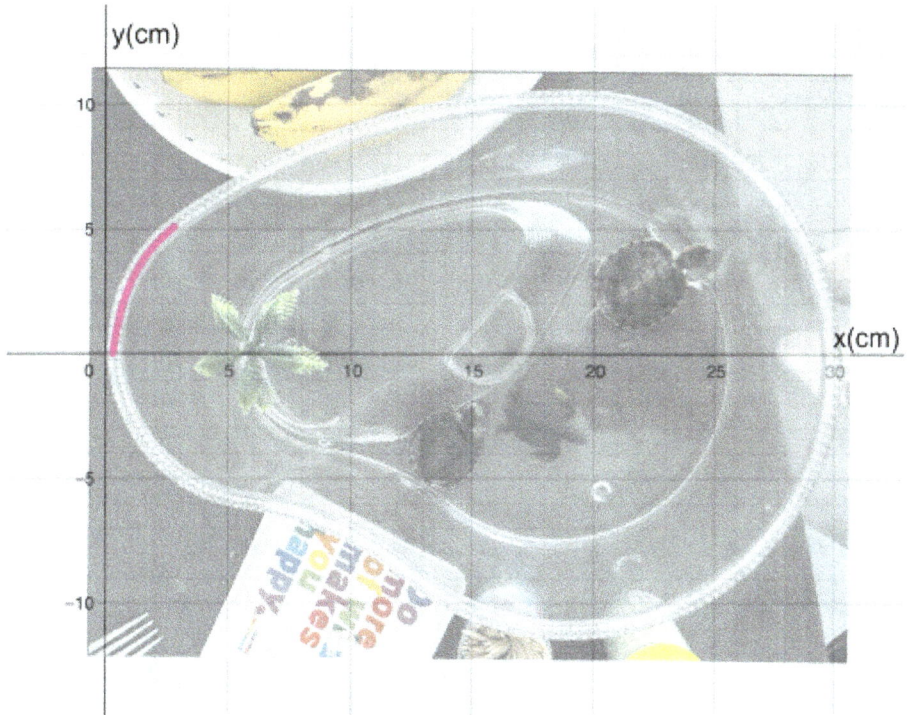

Figure 9 - Log function

This graph is a log function shown in the equation:

$$y = 2.857 + 2.2185 \ln(x), \; 0.280 < x < 2.761$$

The area under this curve was given by,

$$\text{Area 3} = \int_{0.280}^{2.761} (2.857 + 2.2185 \ln(x))\,dx$$

Using my calculator,

$$\text{Area 3} = 8.645 \; cm^2$$

For the second section of the enclosure, I had to decide what function would fit this area well. I tried a quadratic function as it looked like part of a parabola. This fit well. It can be seen in Figure 10 below.

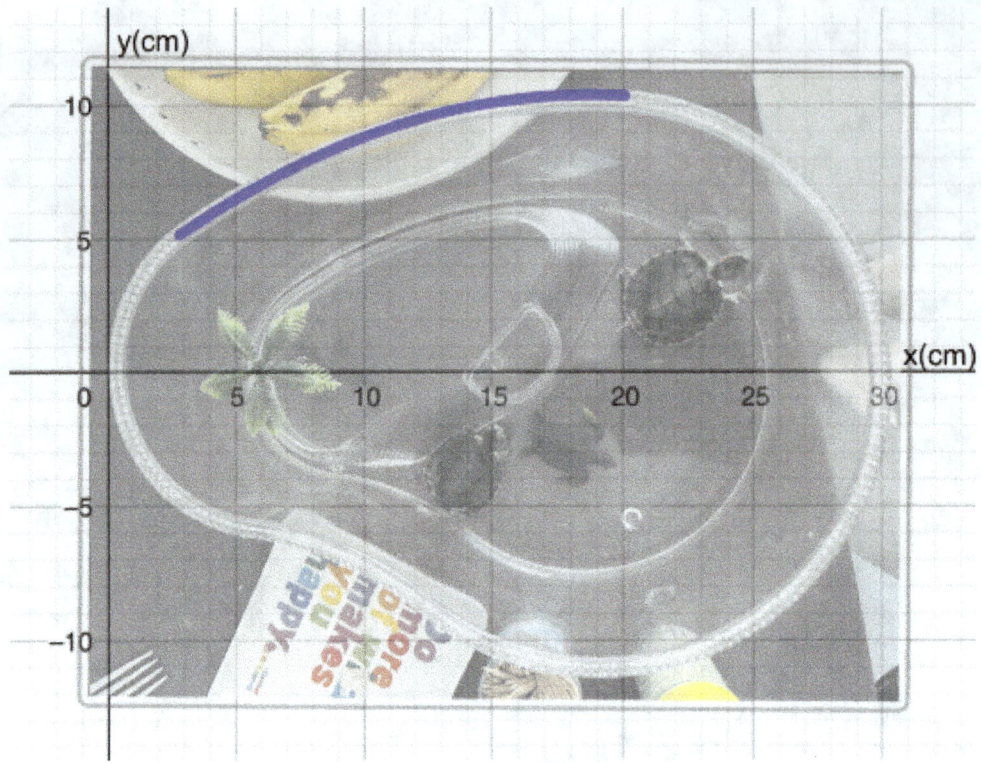

Figure 10 - Quadratic function

The above graph is given by the equation,

$$y = -0.020309x^2 + 0.76498x + 3.1997$$

I limited the domain to $(2.7861 < x < 20.071)$

Then I integrated the function,

$$\text{Area 4} = \int_{2.7861}^{20.071} \left(-0.020309x^2 + 0.76498x + 3.1997\right)dx$$

$$= \left[-0.006769997x^3 + 0.38249x^2 + 3.1997\right]_{2.7861}^{20.71}$$

I then substituted the limits,

$$\text{Area 4} = \left(-0.006769997(20.071)^3 + 0.38249(20.071)^2 + 3.1997(20.071)\right)$$

$$- \left(-0.006769997(2.7861)^3 + 0.38249(2.786)^2 + 3.1997(2.7861)\right)$$

$$\text{Area 4} = 163.57 - 11.73 = 151.84 \, cm^2$$

For the last section on the bottom half, I thought a quarter circle would fit well here.

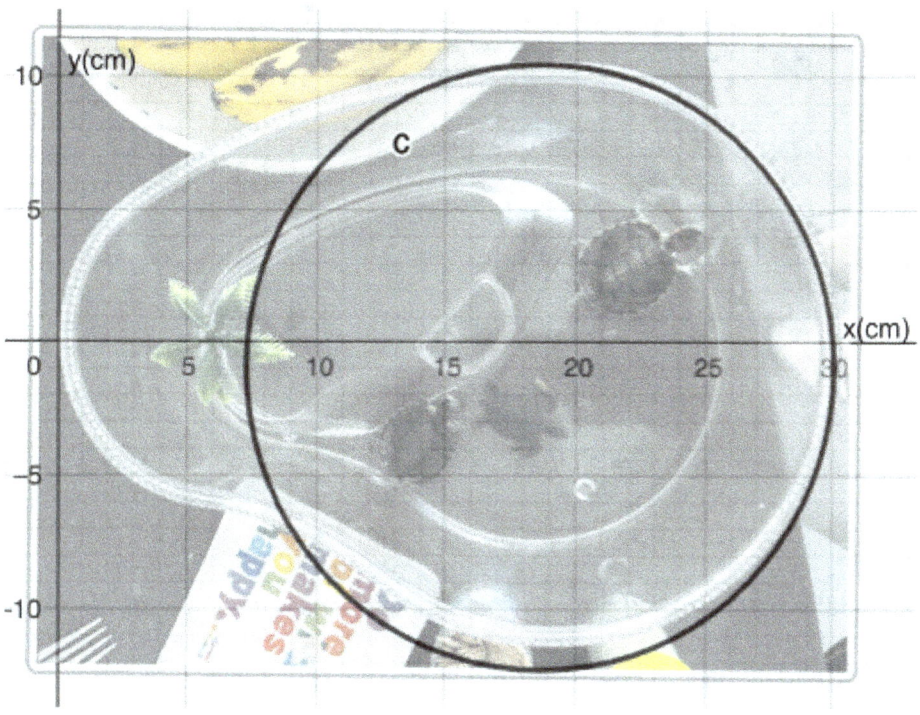

Figure 11 - Circle

As you can see, the circle fit well in the area I needed it to. I limited the domain to $(20 < x < 30)$, and this is shown in Figure 12 below.

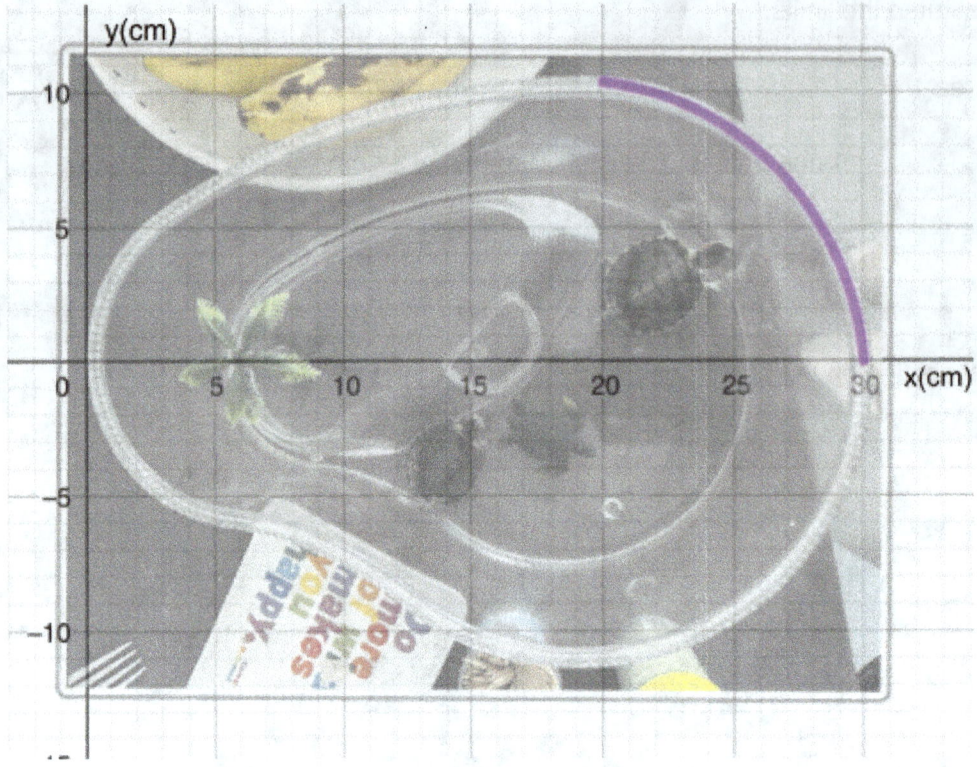

Figure 12 - Quarter circle

The equation of the circle from Figure 11 was

$$(x - 18.651)^2 + (y - 0.92558)^2 = 129.65$$

I again rearranged the equation of a circle to find a function that I could integrate. The following equation gives the purple graph in Figure 12 with the domain limited.

$$y = -0.92558 + \sqrt{129.65 - (x - 18.651)^2},\ 20 < x < 30$$

$$\text{Area 5} = \int_{20}^{30} = \left(-0.92558 + \sqrt{129.65 - (x - 18.651)^2}\right) dx$$

$$\text{Area 5} = 77.22$$

To calculate the total area of the bottom half, I need to find the sum of all 3 areas calculated,

$$\text{Area 3} + \text{Area 4} + \text{Area 5} = \text{Lower Area}$$

$$8.645 + 151.84 + 77.22 = 237.705\ cm^2$$

To find the total area, all I needed to do was to add the upper area to the lower area.

Upper Area + Lower Area = Total Area of Enclosure

$$232.64 + 237.705 = 470.345 \ cm^2$$

The total area for my turtle enclosure is $470.345 \ cm^2$

Finally, I had the answer to my question. I went to see my vet to see if this was sufficient space for my turtles to live and thrive. The vet gave me a direct answer - No. He said I needed more space. I then went and bought a new house for my turtles, which you can see in Figure 13 below. Now they are much happier. Though happiness is not measurable, the turtles' physical activity has increased, and one of the turtles appears to have almost doubled in size since getting this new enclosure. This is interesting to see because in the past two years of owning them, they have never grown this much.

Figure 13 - The new turtle house

C+ If this isn't Personal engagement, I don't know what is

Conclusion

I aimed to calculate the area of my turtles' house to determine if they had enough space. I fulfilled this aim even though it was not the answer I was hoping for. I was hoping that my turtles did have enough space, but math doesn't lie. On a more positive note, it worked out for the best as my turtles got a new and bigger happy home. Throughout the exploration, I learned a lot, and not just that my turtles were unhappy. I learned how to create mathematical models using Geogebra. I learned how to write the equation of a circle, and I learned about the graphs of different functions, including how changing the degree of a polynomial can affect its shape. I am confident that the area I calculated was a reasonable approximation, but there were limitations in my modelling. For example, when I measured the length of the turtle house, it was not done with perfect precision. Also, the functions I used to model the house did not fit the outline of the house perfectly. Having said that, they were close enough to answer my initial question. Overall, I am happy with the exploration and how it turned out. Next time I get a pet, I will ensure I do more research about what kind of surroundings they will need.

Excellent IA

IA Sample 5 Examiner Comments: Cake Optimisation

This is an excellent example of an optimisation IA. The student designs a cake so that the amount of icing is minimized.

Appropriate for: AAHL, AASL, AIHL, AISL

Criterion	SL	HL
A	4	4
B	4	4
C	2	2
D	2	2
E	6	4
Total	18	16

A: This is a very easy read. It is coherent. A peer with good knowledge should be able to easily follow what's happening. It is very well organized. It flows from one section to the next and it is concise. Short and sweet with no extra unnecessary maths.

B: Very little to complain about here. All variables are defined. The mathematical language is correct. The equations look good in the centre with clear descriptions beside them. The student missed an approximately equal to sign on page 9 which could have been costly, but this was the only error I could find. Rounding is mentioned and explained correctly.

C: There is good personal engagement. It is original and the student did design their own cake. However, there is not enough engagement with the maths to give 3/3. It becomes a bit too much like a textbook problem once the formulas are created.

D: There is good meaningful reflection throughout the IA. Noted on pages 3, 6, 7, 9 and 13. Reflection is good and helps guide the exploration but there is not enough critical reflection to award 3/3.

E: The student clearly understands what they are doing. The level is appropriate to earn the top mark in criterion E for SL but fails to achieve 5 in HL as it lacks some sophistication. To achieve 5 and 6 in HL, the maths needs to be a bit more advanced.

Designing the perfect wedding cake for an Icing hater

Nice Title

Page Count: 13

Introduction

I've always loved baking and being in the kitchen. My mom always worked from home in the kitchen and was constantly cooking meals, so our family enjoyed congregating in the kitchen to spend time with one another. My family also has multiple chefs and restaurant owners in the family. My uncle was a contestant on MasterChef, my aunt owns 3 restaurants and bars in my home country, and my dad's cousin is a world-famous chef who is a judge on Master chef and The Taste, because of this cooking and baking has always intrigued me. With all of this in mind and since my cousin who hates icing is getting married in 2023, I decided to design a wedding cake for her with the least amount of icing possible.

The aim of this exploration is to design a 3-layer wedding cake which minimizes the surface area of the cake using calculus.

Good Intro, rationale and aim

My plan for this exploration was to first design a model cake using 3 cylinders, then use the variables in the model to create a formula and use calculus to find the radius which minimizes the surface area. Using that radius my plan was to then calculate the exact dimensions of the cake.

Designing the Cake

When designing the cake, I took into consideration that my cousin wants a traditional wedding cake with 3 cylindrical layers stacked on top of each other and with equal heights, h. Additionally, for aesthetic purposes the cake will have proportional radii. When I asked my cousin she suggested that the top cylinder should be 3 quarters of the one below. This means that the radii follow a geometric sequence with common ratio ¾. I drew the design on my iPad using r, r_1, and r_2 to represent the radius of each cake layer as shown in Figure 1 below.

Figure 1. Side View of the cake

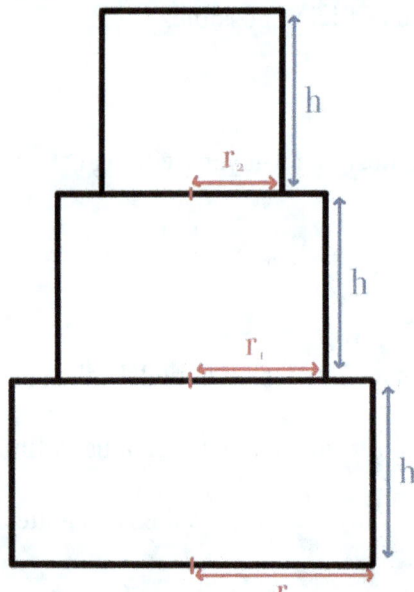

As I wanted all the radii of the cake layers to be written in terms of a common variable, I rewrote the each radius in terms of r, as follows,

$$r = r$$

$$r_1 = \frac{3}{4}r$$

$$r_2 = \frac{3}{4} \times \left(\frac{3}{4}r\right) = \frac{9}{16}r$$

C+ B+ E+ This whole section is very creative and demonstrates a good understanding of the maths that is going to be involved

The volume of the cake was fixed to be 30,000 cm³ as approximately 100 people will be attending the wedding and my dad's cousin Helena she said a volume of wedding cake per person of $300 \, cm^3$.

My original plan was to use only Figure 1. The side view of the cake to calculate both the volume and surface area of the cake, however, I found it difficult to visualize and calculate the top surface area of the cake without the model from above as each cake layer removes some of the top surface area of the layer below it. Additionally, by drawing this model I was able to conclude that the total surface area of the top of each cake layer was equivalent the area of a circle with radius r as seen from Figure 2 because the Green plus Pink plus Blue surface areas are equal to the full area of the bottom circle.

Figure 2. Top View of the cake

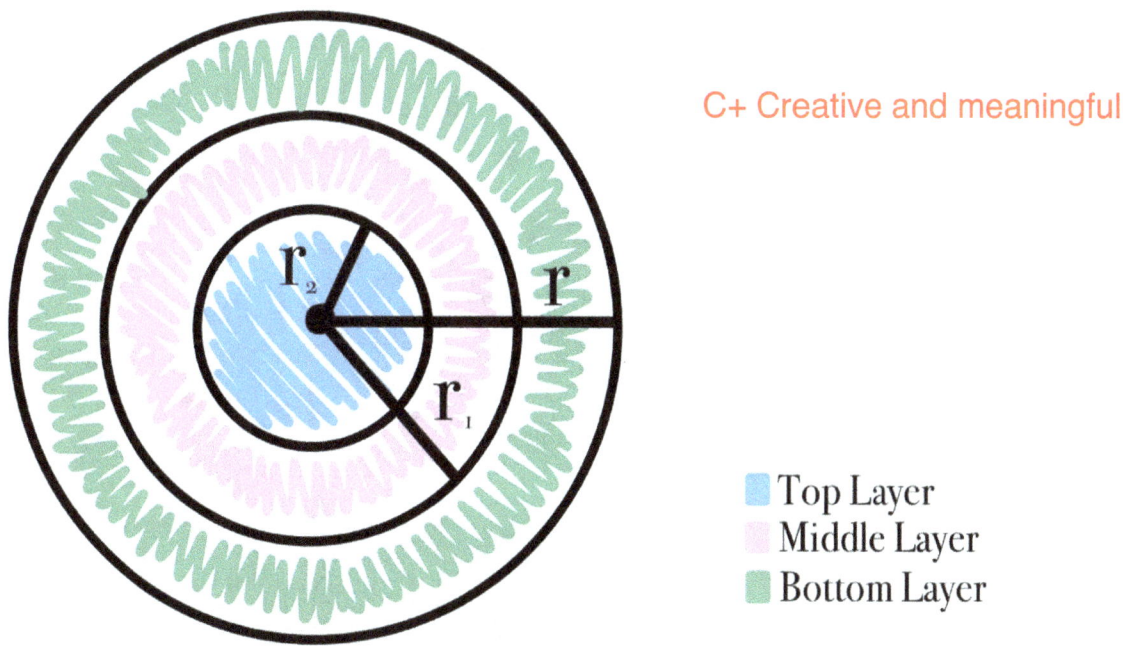

C+ Creative and meaningful

Top Layer
Middle Layer
Bottom Layer

Volume & Surface Area Formula's

The formula used to find the volume, V, of a cylinder is,

$$V = \pi r^2 h$$

The wedding cake I designed consists of three cylinders all with different radii but with equal height. Therefore, the equation for the volume of the first cylinder will be,

$$V = \pi r^2 h$$

the equation for the volume of the second layer will be,

$$V = \pi \left(\frac{3}{4}r\right)^2 h$$

And the equation for the volume of the third layer will be,

$$V = \pi \left(\frac{9}{16}r\right)^2 h$$

B+ E+ This mathematical communication is very good. Understanding demonstrated

This equals 30,000, the volume of the wedding cake.

$$V = \pi r^2 h + \pi \left(\frac{3}{4}r\right)^2 h + \pi \left(\frac{9}{16}r\right)^2 h$$

$$\pi r^2 h + \pi \left(\frac{9}{16} r^2\right) h + \pi \left(\frac{81}{256} r^2\right) h = 30000$$

$$\pi r^2 h \left(1 + \frac{9}{16} + \frac{81}{256}\right) = 30000$$

$$\pi r^2 h \left(\frac{481}{256}\right) = 30000$$

$$\pi r^2 h \left(\frac{481}{256}\right) = 30000$$

Finding the surface area, SA, of the cake was a bit more difficult. The formula for surface area of a cylinder is,

$$SA = 2\pi r h + 2\pi r^2$$

The formula above has 2 parts, the first, $2\pi rh$, shows the curved surface area of a cylinder and the second, $2\pi r^2$, shows the top and bottom surface area of a cylinder. I firstly found the curved surface area, CSA, of each layer. The bottom layer was,

$$CSA = 2\pi rh$$

the middle layer was,

$$CSA = 2\pi \left(\frac{3}{4}r\right) h$$

and the top layer was,

$$CSA = 2\pi \left(\frac{9}{16}r\right) h$$

To find the surface area of top of the cake initially seemed difficult as each new layer removed part of the top of the below layer. However, as previously noted from Figure 2. The Top View that if I added all the remaining surface areas which were the top surface areas, *TSA*, I had a complete circle with radius r which allowed me to use the area of a circle formula to find the surface area of the top of the cake.

$$A = TSA = \pi r^2$$

Lastly, it is important to remember that the surface area of the bottom of the wedding cake can be neglected or assumed to be zero as bakers do not put icing on the bottom of the cake.

D+ good reflection

Adding together all the surface areas gave me the following equation.

$$SA = 2\pi rh + 2\pi \left(\frac{3}{4}r\right)h + 2\pi \left(\frac{9}{16}r\right)h + \pi r^2$$

$$= 2\pi rh + \frac{6}{4}\pi rh + \frac{9}{8}\pi rh + \pi r^2$$

$$= \pi rh \left(2 + \frac{6}{4} + \frac{9}{8}\right) + \pi r^2$$

$$SA = \frac{37}{8}\pi rh + \pi r^2$$

Minimizing the Surface Area

In order to find the radius that minimizes the surface area, I needed to find the surface area in terms of r only. This meant finding the formula for the height, h, in terms of the radius, r. To find this I used the volume equation found before.

$$30000 = \pi r^2 h \left(\frac{481}{256}\right)$$

Rearranging,

$$h = \frac{256}{481} \times \frac{30000}{\pi r^2}$$

Up until now I was able to avoid rounding errors by keeping all equations in exact value or fraction form and not converting to decimal. I substituted the formula found for the height, h, into the Surface area equation.

$$SA = \frac{37}{8} \pi r h + \pi r^2$$

Substituting,

$$SA = \frac{37}{8} \pi r \left(\frac{256}{481} \times \frac{30000}{\pi r^2}\right) + \pi r^2$$

Then simplifying,

$$SA = \frac{960000}{13r} + \pi r^2$$

$$SA = \pi r^2 + \frac{960000}{13r}$$

My next step was to find the value of r which minimizes this surface area function. To do this I needed to find the derivative of the function. However, in order to differentiate, I had to rewrite the formula for surface area without variables in the denominator and negative indices where required.

$$SA = \pi r^2 + \frac{960000}{13}r^{-1}$$

Differentiating with respect to r,

$$\frac{d(SA)}{dr} = 2\pi r - \frac{960000}{13}r^{-2}$$

To find the minimum we equate the derivative to zero, this is because the rate of change at a turning point is equal to zero.

$$2\pi r - \frac{960000}{13}r^{-2} = 0$$

Rearranging and solving,

$$2\pi r = \frac{960000}{13r^2}$$

$$26\pi r^3 = 960000$$

$$r = \sqrt[3]{\frac{960000}{26\pi}}$$

$r \approx 22.736$ (3 decimal places)

22.7 (1 decimal place)

This is maybe the most important result in my exploration as it tells me that the bottom layer should have a radius of 22.7 cm and the remaining layers can be easily calculated using the common ratio. I rounded to one decimal place as in the real-world bakers are unlikely to be as precise as we are in mathematics and the baker will not need more decimal places than 1. In order to be precise and accurate with my future calculations I also kept the result to 3 decimal place. Additionally, while this is the value of r, that minimizes surface area, I also wanted to find out what this surface area would be. By substituting this value of r into the original formula for surface area I found the minimum surface area of the full cake.

Well written. Good understanding demonstrated and good reflection.

$$SA = \pi r^2 + \frac{960000}{13r}$$

$$SA = \pi(22.736)^2 + \frac{960000}{13(22.736)}$$

$$SA = 4871.954$$

$$SA = 4872$$

My final surface area here I rounded to the nearest whole integer as this is sufficiently accurate for the remainder of my calculations. At this point, I realized that I wasn't sure if I had found the value of r which would minimize the surface area or maximize it. Although I could be sure I found where the derivative was equal to zero the only way to confirm that 4871.954 was the minimum was to use the second derivative. This is because when the function is at its minimum point the second derivative will be greater than zero and when the function is at its maximum point the second derivative will be less than zero. To find the second derivative I differentiated the first derivative.

$$\frac{d(SA)}{dr} = 2\pi r - \frac{960000}{13} r^{-2}$$

$$\frac{d^2(SA)}{dr^2} = 2\pi + 2\left(\frac{960000}{13}\right) r^{-3}$$

Simplifying,

B+ E+ More good presentation and understanding

$$\frac{d^2(SA)}{dr^2} = 2\pi + \frac{1920000}{13r^3}$$

Substituting $r = 22.736$,

$$= 2\pi + \frac{1920000}{13(22.736)^3}$$

$$\approx 18.85$$

Therefore, as $18.85 > 0$, $(22.736, 4871.954)$ is a minimum. This can also be confirmed by the graph below of the original Surface Area function, $SA = \pi r^2 + \frac{960000}{13} r^{-1}$. The minimum point is clearly identified. The graph to the left of $x = 0$ can be ignored as the radius must be positive.

Figure 3. Desmos Graph of the Wedding Cakes Surface area against radius r

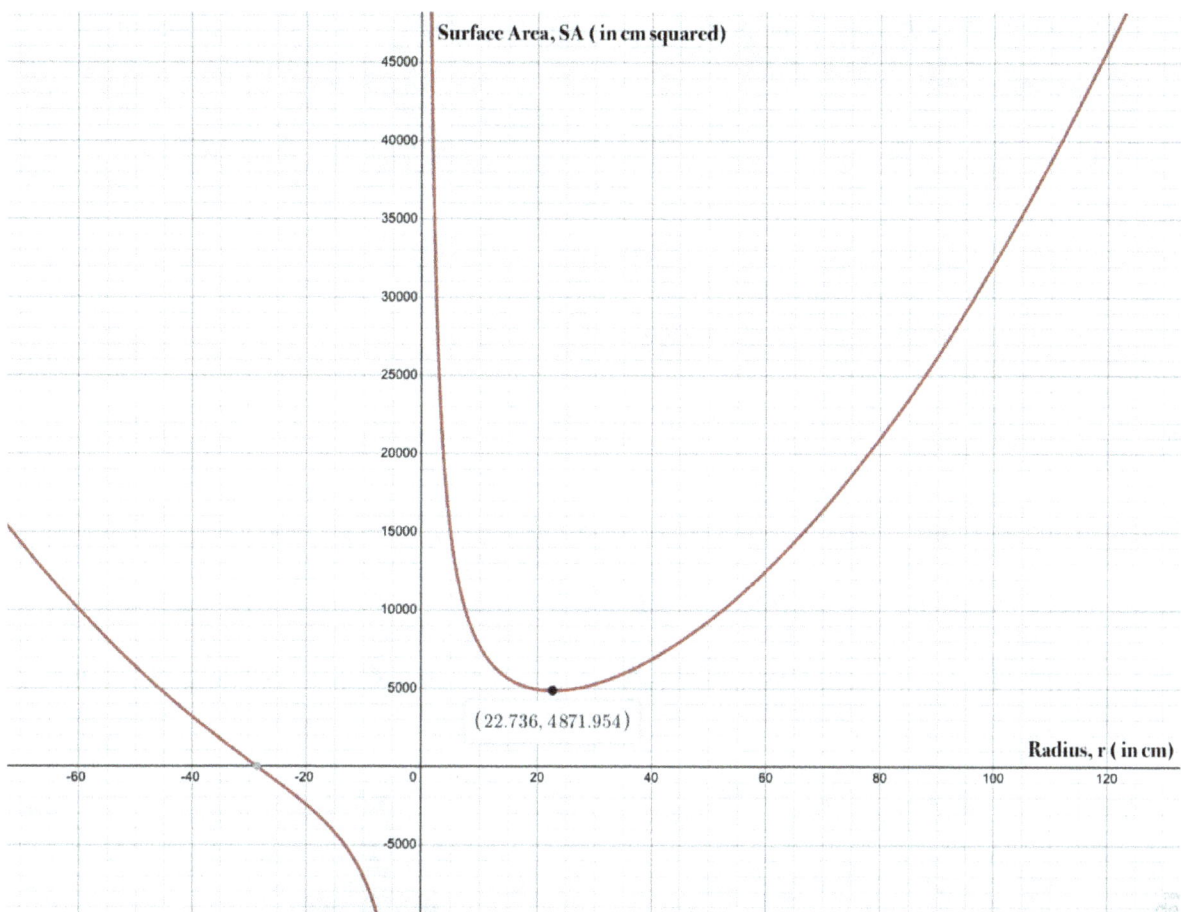

Now that I could be sure that 22.7 was the minimum r, I substituted the 22.7 in as the radius, r, in the height formula to get the height of the cake layers.

$$h = \frac{256}{481} \times \frac{30000}{\pi r^2}$$

$$h = \frac{256}{481} \times \frac{30000}{\pi (22.7)^2}$$

$$h \approx 9.9 \text{ cm}$$

As each layer is 9.9 cm in height this would make the total height 29.7 cm.

Next as the backer will start from the bottom rounded cake layer, I also found the values for the radii of the other 2 layers using the proportions made in the beginning of the exploration.

$$r_1 = \frac{3}{4}r = \frac{3}{4} \times (22.7)$$

$$r_1 \approx 17.0 \text{ cm}$$

$$r_2 = \frac{9}{16}r = \frac{9}{16} \times (22.7)$$

$$r_2 \approx 12.8 \text{ cm}$$

Lastly, using all of the values found in the exploration above I used SketchUp – a 3D modeling application I learned how to use for this Internal Assessment – to create a to scale model of the 3-layer wedding cake. Through the model it is possible to visualize the dimensions of the model.

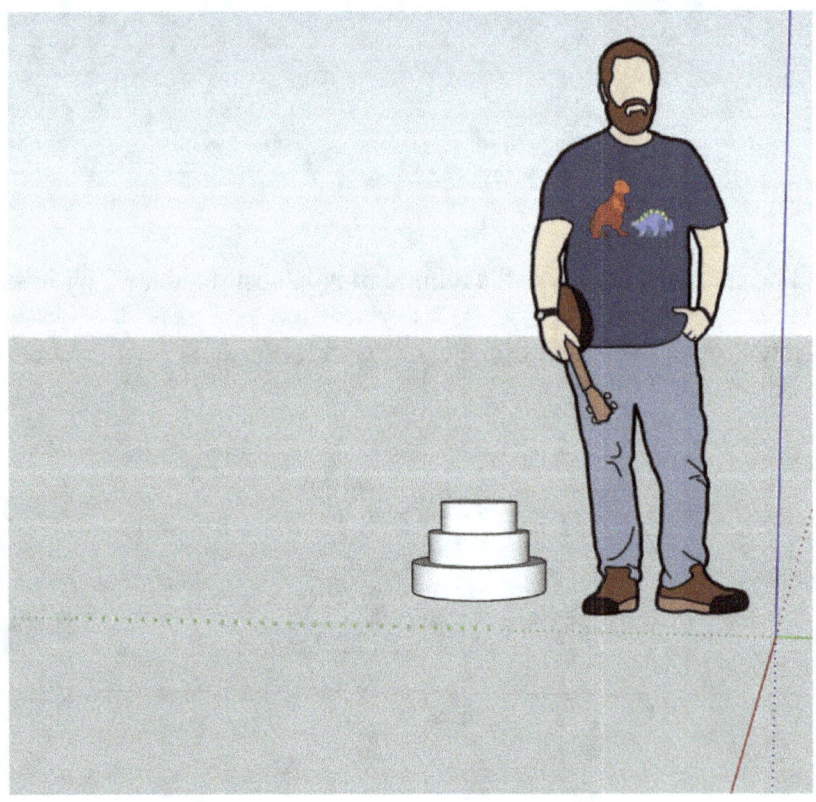

Figure 4. 3D Model of the Wedding Cake

Conclusion

The aim of this investigation, design a 3-layer wedding cake which minimizes the surface area of the cake using calculus, was achieved. I did this by firstly finding the radius that minimized the surface area to be 22.7 cm and the surface area to be 4871.954 cm². Then, by graphing the surface area function and through using the second derivative test I was able to prove that I had found the minimum point. Using the radius, I was also able find the required height of each cake layer, h, to be 9.9 cm and the value of each of the radii. Now, I am fully prepared to take my findings and proportions to the baker and make my cousins wedding cake.

Throughout the investigation, there were many opportunities for errors to be made as the math was all done by hand, however, to minimize errors or miscalculations due to rounding I left all values in exact form rather than convert them to decimal. Additionally, it is important to note that Figure 1 and Figure 2 used to identify my variables and complete my calculations was also drawn by hand as such they are not drawn to scale. Moreover, bakers often use standardized baking sheets so while the dimensions found would minimize the surface area of the cake it may not be feasible to implement these new dimensions in the real world. Furthermore, wedding cakes typically have designs such as flowers made of icing on them, so it is important to note that the value for the minimum surface area or in other words minimum icing needed for the cake does not account for any designs on the wedding cake.

At a later date, an extension of this project could be done to minimize the surface areas of non-cylindrical or different shaped cakes. Additionally, this method could be used to minimize packaging needed to transport and store not just cakes but other products. To conclude, it is possible to use calculus to minimize the surface area of a 3-layer wedding cake.

IA Sample 6 Examiner Comments: Swimming Normal Distribution

This is a well-done IA where a student uses normal distribution to find the probability of reaching the Youth Olympics. I gave this 18/20 and the IB increased it to 19/20.

Appropriate for: AASL, AISL, AAHL, AIHL

Criterion	SL	HL
A	4	4
B	3	3
C	3	3
D	3	3
E	6	4
Total	19	17

A: The IA is very clearly structured, guiding the reader from motivation to data, to model, and to interpretation. Headings are clear and consistent, and the overall flow of the IA supports understanding. Diagrams (e.g. normal curve illustrations, probability regions, data tables) are relevant and well-placed. The layout supports comprehension without clutter. Lots of subheadings are not recommended but works well here.

B: Functions are written clearly, and the notation is mostly consistent. Mathematical notation is correct and appropriate throughout, especially in defining the normal distribution model, z-scores, and probability calculations. Diagrams and symbols are used well. Stars for the normal distribution is not appropriate. A curve should have been used.

C: This is clearly a highly personal topic. The student's own swimming goal (qualifying for the Youth Olympics) provides an authentic and engaging motivation. The choice to use real data and to model performance probabilistically shows initiative. The extension where the student calculates what average time she'd need to increase her qualification probability is especially strong and shows ownership.

D: There is regular critical reflection throughout the IA. The Reflection sections are not normally recommended but it is acceptable here and makes the reflection very clear. There's thoughtful reflection in the second part: estimating the required new mean time and discussing whether that's realistically achievable. The limitations of using a normal distribution for time data are acknowledged, including the fact that there is not a lot of data.

E: The student uses the normal distribution appropriately: defining the model, finding probabilities, computing z-scores, and using inverse normal techniques. The mathematical reasoning is accurate and well explained. The extension to calculate a new required mean based on desired probability is a subtle and excellent use of inverse distribution and algebraic reasoning. Thorough understanding is demonstrated hence 6/6 for SL.

Title could be a lot better. Why not "Can I win the Olympics?"

Plotting the normal distribution with swimming times
IB Mathematics SL Analysis and approaches
Internal Assessment

May 2021
Page number: 13

Introduction:

The aim of this exploration is to find the probability that I have to qualify for the most important competitive swimming meet for my age group which is, the youth Olympic games. I will do this by investigating my average swimming time for my 50 meter freestyle event in comparison to the qualifying swimming time for 50 meter freestyle for my age group, through the probabilities and graphing of normal distribution. This will allow me to assess if participating in the youth Olympic games is an achievable personal target. I also aim to find the probability of qualifying for another event which requires a different qualifying time. Overall, this will allow me to compare both and see which event I should enter according to the higher probability of me qualifying.

I have chosen this topic because I have an interest in swimming and have been tracking my development all through my swimming path. Ever since I started swimming competitively at the age 8, I have participated in many development galas and swimming team meets. By completing all these competitions, I have been able to improve my times over the years and reach my personal best times today.

To fulfill my aim, I will collect all the swimming times for the 50 meter freestyle event I have done in the past 3 years. I will then find the mean and standard deviation of all my times and plot them as a normal distribution curve. I will then determine if the average time is similar to the qualifying times for my particular age group for an event. I will model my times using a normal distribution.

Good intro. Aim, plan, rationale. I am now interested. Photos add.

Data collected

I have gathered my swimming times for the 50 meter freestyle events that I have participated in during the last three years. I have organised the times into a table with the date of the competition the time was collected at.

Table to show my 50 meter freestyle competition times.

Date (dd/mm/yyyy)	Time (seconds)
29.10.2016	30.23
28.4.2017	29.62
28.4.2017	29.75
14.10.2017	30.85
23.4.2018	31.34
8.12.2018	29.25
4.5.2019	29.73
22.6.2019	29.16
27.9.2019	29.27
18.10.2019	28.69
29.11.2019	28.70

By using this data above, I will find the mean as well as the standard deviation related to my results in order to make a normal distribution graph. This will allow me to find and look at the probabilities I have in order to reach certain events.

Reflection 1

All of the data I gathered was extracted from a swimming database called meet mobile that keeps records of each swimmer's times from different competitions. However, some of these times were recorded in a short course pool, and others in a long course pool. Short course pools are 25 meters long, while long course pools are 50 meters long. This makes a difference in swimming times. Times taken in long course pools are often slightly higher because it requires more stamina as you are not able to take advantage of the tumble turn impulsion on the wall during 50 meter races. However, in short course pools, times will be faster due to being able to push off the wall and underwaters before breaking to the surface of the water. Thus, that is a limitation of my data, as not all times represented were taken from the same length of swimming pool.

D+ Clear reflection

Finding the mean

To calculate the mean of my data set, I have taken the sum of all my data points and divided it by the number of data points in my data set.

μ = represents the population mean (seconds)
Σ = represents the sum of all data points in the population
x = swimming times (seconds)
n = represents the total number of data points in the population

$$\mu = \frac{\Sigma x}{n}$$

$$\mu = \frac{30.23 + 29.62 + 29.75 + 30.85 + 31.34 + 29.25 + 29.73 + 29.16 + 29.27 + 28.69 + 28.70}{11}$$

$$\mu = 29.69 \text{ seconds}$$

Finding the standard deviation (SD)

To calculate the standard deviation, I used the mean of my data set previously calculated. Then, for each of my data points, I had to subtract the mean, and square the results. Then I found the mean of all the squared differences. This resulted in the variance. To get the standard deviation from this, I took the square root of the variance. The standard deviation is a measure of how spread out the values of the data set are.

σ = represents the population standard deviation
Σ = represents the sum of all the squared differences
x = represents each single swimming time (seconds)
$\bar{x}$ = represents the mean of the population (seconds)
n = represents the total number of data points in the population

Mean and SD could have been done [on] GDC but this doesn't take from the IA. Presentation is good.

$$\sigma = \sqrt{\frac{\Sigma(x-\bar{x})^2}{n}}$$

$$\sigma = \sqrt{\frac{(30.23 - 29.69)^2 + (29.62 - 29.69)^2 + (29.75 - 29.69)^2 + \cdots}{11}}$$

$$\sigma = \sqrt{\frac{(0.54)^2 + (-0.07)^2 + (0.06)^2 + \cdots}{11}}$$

$$= 0.79777$$
$$= 0.80$$

Reflection 2

I have rounded the mean and standard deviation to two decimal places. In swimming competitions, times are always recorded to two decimal points, as stated by the International Swimming Federation (FINA), indicating seconds and milliseconds. This is done in order to be able to distinguish and separate swimmers in the rankings. Swimming times are continuous data; therefore it does not mean that swimmers have achieved those exact times, but rather this is the closest meaningful number that indicates their swimming time, in accordance with FINA rules for standardization.

These reflection sections are not recommended but they do make the reflection very clear.

Plotting the normal distribution curve

Swimming times generally follow a normal distribution. Therefore, I will use the normal distribution function. To plot the normal distribution curve, I used the excel software from Microsoft. By using my mean and standard deviation, I was able to create the normal distribution curve representing my performance in the last three years. I created a table including the standard deviation increments (-3 to 3), swimming times, and probabilities. The area under the curve is 1, and any shaded area under the curve can represent a cumulative probability.

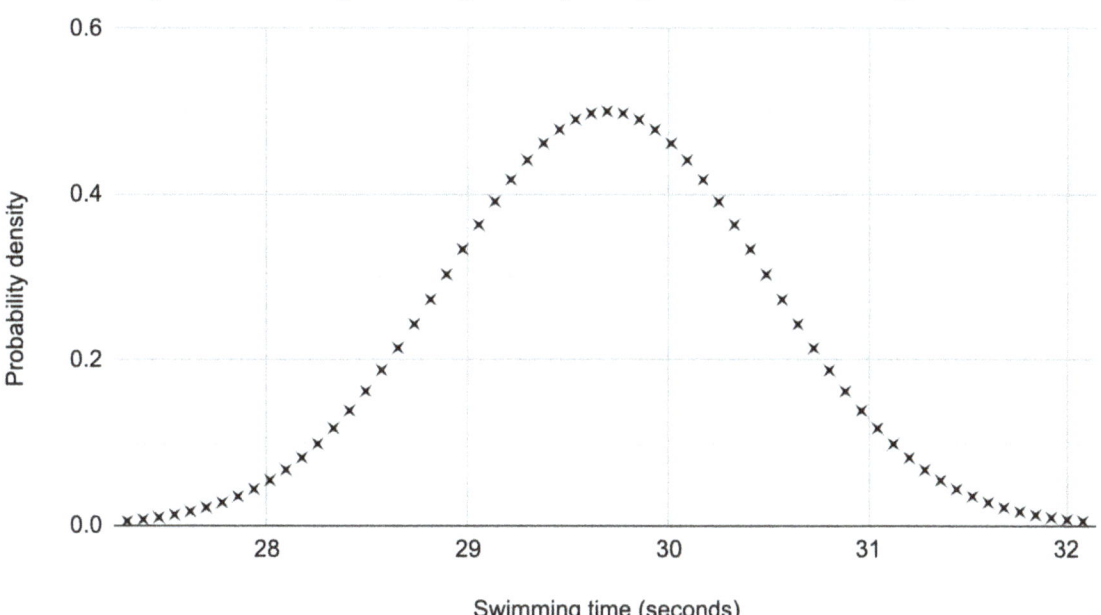

Graph to show the probability density of my 50 meter swimming times.

Not sure why these stars were used. A normal curve would be better.

Reflection 3

The number of data points is not ideal for using the normal distribution function. An ideal sample would consist of a minimum of 30 data points. However, over the past 3 years of swimming, I have only been able to collect 11 times in competition settings. If I collected more data from more previous years, it would skew my data and not give me realistic probabilities that are relevant to me today. By adding times from more than 3 years ago, my mean would increase a lot due to my drastic improvement along the years. This would also increase my standard deviation and my normal distribution curve would not have given me an accurate representation of my current level.

In addition to these data points, collecting data during training sessions or in a pool on my own would also skew my data. They would not be representative of my abilities due to them not being taken in competition conditions. If half my data was from competitions and the other half was from training sessions, it would not properly represent my abilities in a competition environment. Therefore, I decided to calculate my normal distribution probabilities while only using competition times in the past 3 years.

Finding the probability of reaching a qualifying time for the youth olympics

My current mean of swimming times is 29.69 seconds. The qualifying time for the youth Olympic games for my age group for the 50 meter freestyle is 27.39 seconds. Using my graphing calculator, I was able to calculate the probabilities by using the normal CDF function by inputting my mean, standard deviation, and area.

X = represents the random variable that is normally distributed
μ = represents the mean
σ^2 = represents the variance

$$X \sim N(\mu, \sigma^2)$$

$$X \sim N(29.69, 0.80^2)$$

$$P(X < 27.39) = 0.00202$$

$$= 0.20\%$$

B+ E+ Presentation is clear. Use of GDC to find the probability is appropriate here

Graph to show the probability density of my 50 meter swimming times.

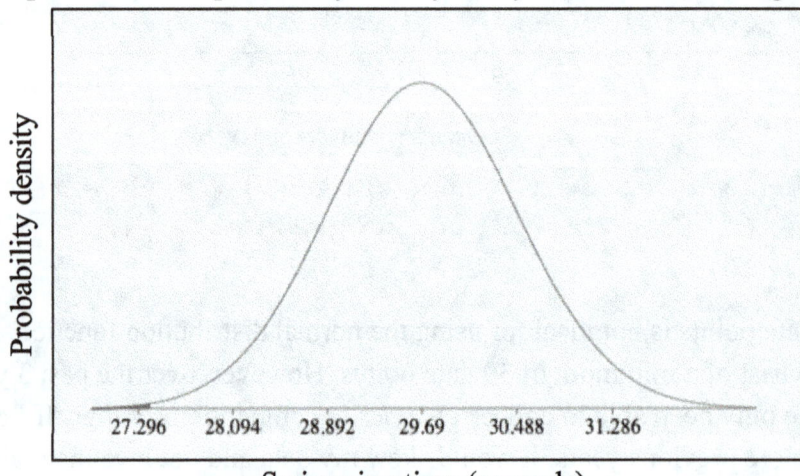

Reflection 4

The probability of me reaching the qualifying time with my level is very slim at the moment. The shaded region of the normal distribution curve indicates the probability of qualifying with my current mean and standard deviation. If I wish to qualify for this event, my overall mean must decrease by a few seconds. In order for this qualifying time to be within my reach reasonably, I must see what my mean should be if I want my probability to increase.

What is the time I can achieve with a 25% probability based of my current performance for the 50m freestyle?

Participating in the youth Olympics for swimming is the aim of all highly competitive teenage swimmers. I understand that it would be a great achievement to qualify. I have chosen the probability of me qualifying to be 25% as I feel like this probability will be realistic to achieve with hard work. If the probability is lowered, it does not give me a good chance of qualifying, meaning it is not worthwhile for me to find out a new mean. On the other hand, if I were to choose a stronger probability, it would be much more difficult to reach the new mean.

I used the inverse normal function on my calculator. By inserting the mean and standard deviation, along with the area of probability, I was able to find the time I have a 25 % probability to achieve with my current mean.

P = probability
x = represents the time I am trying to achieve
k = time according to probability (seconds)

$$X \sim N(29.69, 0.80^2)$$

$$P(x < k) = 0.25$$

$$k = 29.1504$$

$$= 29.15 \; seconds$$

Reflection 5

At the moment with my current performances, I have a 25% probability of reaching the time of 29.15 seconds. This is not good enough in order to qualify. To increase my probability to have a 25% chance of qualifying for the 50 meter freestyle youth Olympics, I must improve my overall mean.

What should my average time be if I want a 25% probability of qualifying for the 50m freestyle?

I have to swim under 27.39 seconds in order to qualify for the event. Assuming my standard deviation stays the same, I must find the new mean time giving me 25% chance of achieving the qualifying time. I will first need to find the z-score in order to know the new mean I must have to achieve that probability.

z-score formula

The z-score indicates how far from the mean a data point is. It measures how many standard deviations below or above the mean a data point is. The z-score can be plotted on the normal distribution, and can range from a value of -3 standard deviations to +3 standard deviations.

z = the z-score
x = represents the value to be standardized
μ = represents the population mean
σ = represents the population standard deviation

C+ D+ this is a really good example of letting the maths guide the exploration and then trying something creative

$$z = \frac{x - \mu}{\sigma}$$

To find the z-score, I used the inverse normal function on the calculator by inputting the area of probability I am trying to find (0.25), 0 as the mean (μ), and 1 as the standard deviation (σ).

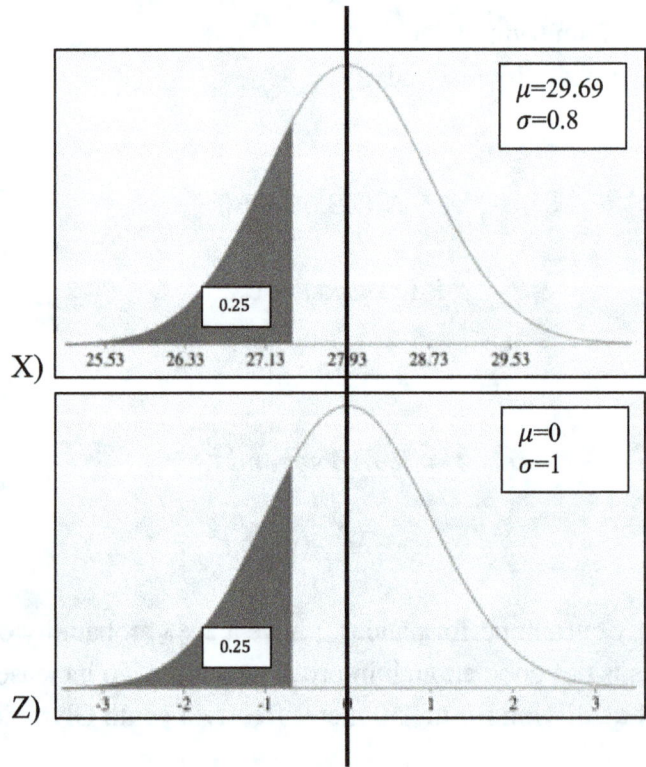

The diagrams illustrate how the z-score can be found. The graph labelled Z can be compared to the X graph, with my appropriate mean of 29.69 seconds, and standard deviation of 0.8. By finding the z-score with a mean of 0 and a standard deviation of 1, it can be applied to my data. The shaded area represents the probability density regarding the standard deviation. Using the inverse normal to find the z-score, I know that the time I want to achieve is -0.67449 standard deviations away from the mean.

$$z = -0.67449$$

$$-0.67449 = \frac{27.39 - \mu}{0.80}$$

$$\mu = 27.9296$$

$$= 27.93 \; seconds$$

Using the new obtained mean while still using the previous standard deviation, I have been able to plot a new normal distribution curve. The shaded area under the curve indicates the probability I have of swimming under the qualifying time of 27.39 seconds.

Graph to show the probability density of my 50 meter swimming times.

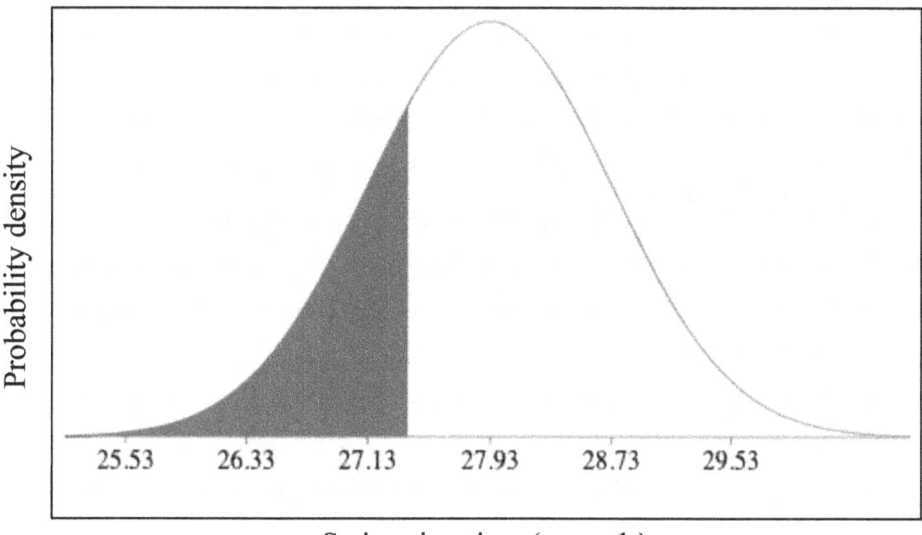

Swimming time (seconds)

Reflection 6
If my training increases and I am able to reach this new mean I will have a strong chance of qualifying. However, this does not seem within my reach at this stage as I have not swam at that time in any competition. Since my personal best time is 28.69 seconds and knowing my personal abilities, I do not have a strong chance of reaching this time either.

$$X \sim N(29.69, 0.80^2)$$

$$(X < 27.93) = 0.013903$$

$$= 1.39\%$$

The probability of even reaching the new mean is very low. By finding these probabilities, I have realized that I was too ambitious as it is not very likely that I will be able to qualify for the youth Olympics 50 meter freestyle. It would take a lot of training to improve my times to reach the required level. Therefore, I must think of qualifying for a smaller scale competition as this will improve my probability. It would be less difficult to achieve the qualifying time. As my probability of qualifying increases, it also improves my probability for success in the actual race when I participate in it.

Finding the probability of reaching a qualifying time for the Middle East Open Championships

My current mean of swimming times is 29.69 seconds. The qualifying time for the Middle East Open Championships for the 16 years and over age group for the 50 meter freestyle is 29.60 seconds. Using my graphing calculator, I was able to calculate the probabilities by using the normal CDF function by inputting my mean, standard deviation, and area.

$$X \sim N(29.69, 0.80^2)$$

$$P(X < 29.60) = 0.455213$$

$$= 45.52\%$$

Graph to show the probability density of my 50 meter swimming times.

[Graph: normal distribution curve with x-axis labeled "Swimming time (seconds)" with values 27.296, 28.094, 28.892, 29.69, 30.488, 31.286, and y-axis labeled "Probability density". The area under the curve to the left of 29.69 is shaded.]

At the moment with my current mean, I have a 45.52% probability of reaching the time for qualifying for the event in the Middle East Open Championships. This is a strong probability and the time I have to achieve to qualify is not far away from this current mean time. Therefore, I will explore the probabilities involved with placing in the event.

What is the probability of me reaching the final of the 50m freestyle event for the Middle East Open Championships

By looking at the results from the 45 girl participants (16 and over) from the 50 meter freestyle of this competition in February 2020, I will find the probability of placing in the top 10 in the finals. The times of all 45 swimmers ranged from the fastest time of 25.63 seconds, until the slowest time of 29.55 seconds. In 10th place was the time of 26.94 seconds, therefore if I wish to place in the top 10, I must go below this time.

$$X \sim N(29.69, 0.80^2)$$

$$P(X < 26.94) = 0.000294$$

$$= 0.03\%$$

This probability is still quite low and therefore does not seem achievable with my current average time. Placing in the top 10 swimmers in the middle east for my age group would be a very high achievement. However, it does not seem realistic to me and I will change aim to be in the 20th place for the rankings in the finals. In 20th place was the time of 27.69 seconds, therefore if I wish to place in the top 20, I must go below this time.

> It is supposed to be double spacing but still very well written. I am finding this very easy to follow.

$$X \sim N(29.69, 0.80^2)$$

$$P(X < 27.69) = 0.00621$$

$$= 0.62\%$$

The probability I have to place in the top 20 is still drastically low. By using the inverse normal function on my calculator, I will find the best time I can achieve with a 25% probability, based on my current mean and standard deviation. I will then link this back to the results of the final event and see in what place I would be able to rank.

$$X \sim N(29.69, 0.80^2)$$

$$P(x < k) = 0.25$$

$$k = 29.1504$$

$$= 29.15 \; seconds$$

In 37th place was the time of 29.14 seconds. In 38th place was the time of 29.23 seconds. Therefore, with a 25% probability, I would place 38th out of 45 swimmers.

Now, I will find the best time I can achieve with a 10% probability. I will then link this back to the results of the final event and see in what place I would be able to rank.

$$X \sim N(29.69, 0.80^2)$$

$$P(x < k) = 0.10$$

$$k = 28.6648$$

$$= 28.66 \; seconds$$

In 33rd place was the time of 28.53 seconds. In 34th place was the time of 28.77 seconds. Therefore, with a 10% probability, I would place 34th out of 45 swimmers. Since my personal best time for this event is 28.69 seconds, I have a good chance of achieving this position in the rankings.

Reflection 7
With these particular findings of the results, I now know that I do not have an exceptionally high chance of placing highly in the ranking of the final of this event. The swimmers participating in the event are highly competitive. Therefore, I think participating in the event would allow me to push myself. It would give me more motivation to swim with very fast swimmers and could possibly allow me to make new personal best times.

There is no way to miss that this student has been reflecting

Conclusion

By applying the approach of probability, using the normal distribution model, I have been able to understand my swimming level and define my next target. Based on all the results, I have decided that my target for the swimming season is to continue to participate in many competitions to gather more times. By increasing my training, I will be able to increase my mean swimming time, as well as be able to qualify and participate in the Middle East Open Championships.

Through this exploration, I have been able to achieve my aim to find the probability of qualifying for the Junior Olympics for the 50 meter freestyle event. I did this by collecting all my recent competition swimming times in the last 3 years. This allowed me to draw a normal distribution curve and find out the certain probability densities associated with making certain qualifying times. As I realized that my probability of qualifying for the Junior Olympics was too low, and the mean I had to achieve to have a higher probability was out of my reach, I decided to investigate the probability of qualifying for a lower scale competition. The Middle East Open Championships had a much more achievable qualifying time, as with my mean, I already had a 45.52% probability of qualifying. I therefore investigated the probabilities of placing in the rankings of the event. By comparing my probabilities for these two competitions, I decided it would be more realistic for me to compete in the Middle East Open Championships.

In the future, I will focus on participating in more competitions to make my data more accurate regarding my swimming level and physical state. As I progress, my personal times will get better, therefore lowering my mean and increasing my standard deviation. It would give a better representation of my progressing level. As I continue to train for swimming, I will increase my training session intensities and physical conditioning training. This will allow my goals to be more achievable overall.

Top class IA

IA Sample 7 Examiner Comments: Russian Dolls

This is an excellent HL IA where the student uses modelling and volume of revolution to find the volume of his Russian dolls.

Appropriate for: AAHL, AIHL,

Possible: AASL, AISL (you would need to learn volume of revolution)

Criterion	SL	HL
A	4	4
B	3	3
C	3	3
D	3	3
E	6	5
Total	19	18

A: Exploration is very well written. It is coherent and well organised. It is very easy to follow for a peer. Exploration is concise. Every page is important. There is a clear aim. Introduction and conclusion are present and well done. There is a very nice, easy flow throughout the whole exploration. All graphs and tables and diagrams are in the appropriate place.

B: Generally, the mathematical communication is very good. It is appropriate, relevant and consistent throughout except for one error on page 16 where the ln function is not entirely in the bracket. The mathematics ends up being correct but 1 mark was taken away from this criterion.

C: This is a unique, creative exploration. The personal engagement really comes through when reading. The student is clearly engaged in the maths and their results and thought process guide the exploration forward. Lots of examples noted on the paper. Volume of revolution IAs can be quite common and similar but this one was unique and very interesting to follow. The part where the student used a scale factor to come up with a geometric sequence was particularly pleasing.

D: There is a substantial amount of critical reflection in the exploration. The student reflects on their model and their engagement with the maths drives the exploration. Lots of reflection noted on the paper but some examples mentioned on page 7, 8 and 10.

E: Mathematics is clearly well understood and explained. Mathematics is commensurate with the level of the course. Modelling is done well and student explains why he chose specific models and how he did the fitting. Integration is well done and understood. Rigor and sophistication are demonstrated on multiple occasions.

Introduction

As a child, one of my fondest memories was visiting my grandmother in Kazakhstan who had a beautiful Matryoshka doll set displayed in her kitchen shelf (a replica model seen in Figure 1). I remember being fascinated by how each doll could reveal a smaller and more detailed doll inside with an even smaller and detailed doll within. My grandmother would often store various seasonings such as cinnamon, turmeric, paprika and cumin within these dolls and use them whilst cooking. Seeing this immense storage capacity originating from one fixed volume intrigued me but I could never determine by how much the cumulative volume of the successive dolls superseded the volume of the initial doll. To determine this, the volumes of the dolls must be calculated and compared.

Figure 1: A replica set of my grandmothers Matryoshka dolls

The aim of this investigation is to find the best model to calculate the volume of the largest Matryoshka doll and determine the cumulative volume of the following dolls within. I will then compare the volumes and determine how much more space the smaller dolls possess than the largest, initial doll.

A+ Clear aim

Plan

I will start by obtaining replicas of my grandmother's Matryoshka dolls. This will allow me to measure the diameter and height of the largest doll which can be verified by the measurements my grandmother will take of her own doll. I will then use the verified dimensions of the largest doll in order to model it on GeoGebra and plot graphs modeling the Matryoshka doll accordingly. I can use sinusoidal modeling or a polynomial function to outline the doll and if these methods/functions have some degrees of error to them I can use a combination of a logarithmic function, a cosine function and an elliptical function to produce a presumably more accurate piecewise function modeling the doll.

Clear plan and well written

The volume of the doll can then be calculated using the volume of revolution formula. This is around the x-axis.

$$\text{Volume} = \pi \int_{x_1}^{x_2} (f(x))^2 \, dx$$

Models

The first graph I used to model the Matryoshka doll in Figure 2 was a sin function which was deduced by my knowledge regarding sinusoidal modeling. I chose to use sinusoidal modeling as the doll seemed to have a smooth, symmetric and rounded shape that tapers at the bottom and the top, similar to a sinusoidal function which has alternating wide and narrow segments.

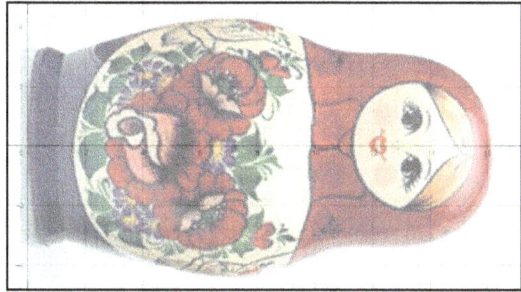

Figure 2: Image of the initial Matryoshka doll

This is seen in the general shape of the doll where the base of the doll in Figure 3 starts at a minimum (where the doll is narrow), reaches a maximum near the middle (where the doll is wider) then tapers off towards the top. C+ good personal engagement. The student is looking forward and thinking about what might make a good model

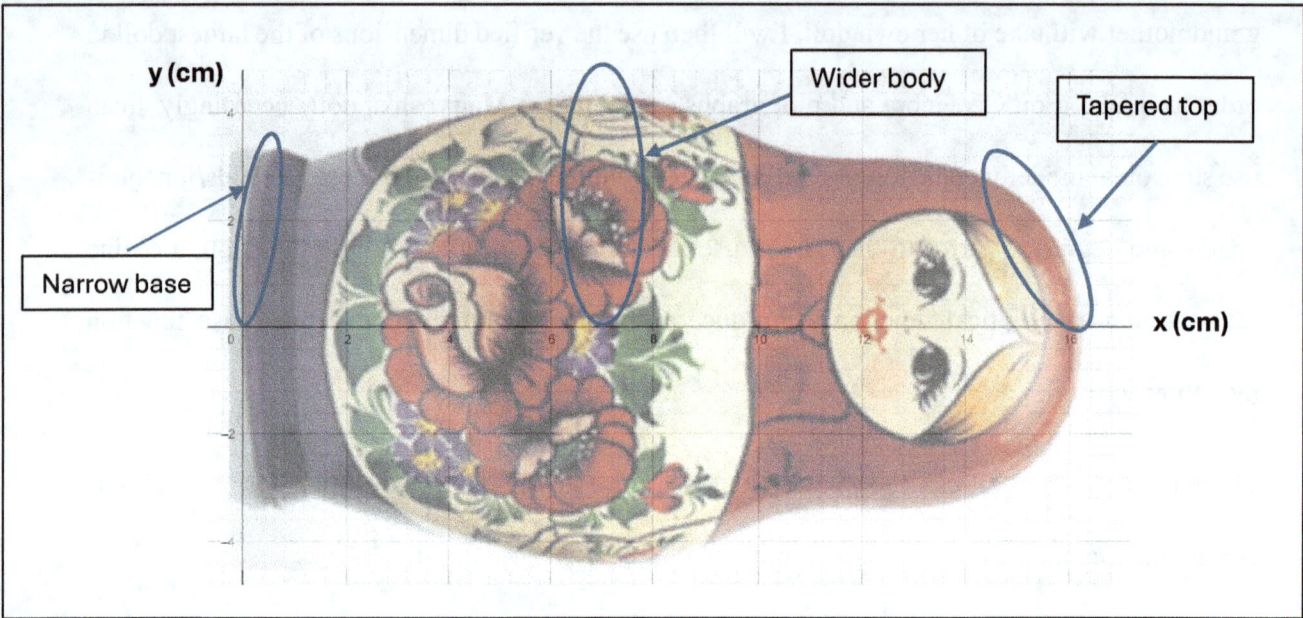

Figure 3: Scaled image of initial Matryoshka doll

The general formula of a sinusoidal curve can be defined as such:

$$f(x) = a\sin(b(x+c)) + d$$

By observing the scaled Matryoshka doll in Figure 3, x-intercepts, maximum and minimum points can be determined to produce values for a, b, c and d. This will form a sinusoidal function modeling the Matryoshka doll where afterwards a volume can be calculated.

In the sinusoidal function a refers to the amplitude and can be calculated as such:

$$a = \frac{\text{maximum}_y - \text{minimum}_y}{2}$$

$$a = \frac{4.3 - 0}{2} = 2.15$$

The period or stretch is the horizontal distance over which one complete sine wave occurs and is represented in the form:

$$\text{Period} = \frac{2\pi}{b}$$

Therefore, by doubling the distance between the maximum and minimum, the period can be determined and b can be calculated.

$$\text{Period} = 2 \times (16.2 - 7) = 18.4$$

$$b = \frac{2\pi}{18.4} = 0.34 \text{ (2 d.p)}$$

The horizontal translation (c) is a measure of how much the sine wave is shifted to the right or left. In a standard sin wave, $y = \sin(x)$, the first maximum occurs at $x = \frac{\pi}{2}$. In the scaled image of the doll, the first maximum occurs at $x = 7$. However, with the modified sine function including the b value, the graph of $y = \sin(0.34x)$ has a maximum at $x = 4.62$. Therefore, the horizontal translation (c) is $(7 - 4.62) = 2.38$

Finally, d represents the vertical shift of the sinusoidal model and can be calculated as follows:

$$d = \frac{\text{Maximum}_y + \text{Minimum}_y}{2}$$

$$d = \frac{4.4 + 0}{2} = 2.2$$

E+ Very good understanding of sinusoidal modelling

Culminating the values found for a, b, c and d produces the sinusoidal function:

$$f(x) = 2.15 \sin(0.34(x - 2.38)) + 2.2$$

Using the newfound function, the following graph was produced:

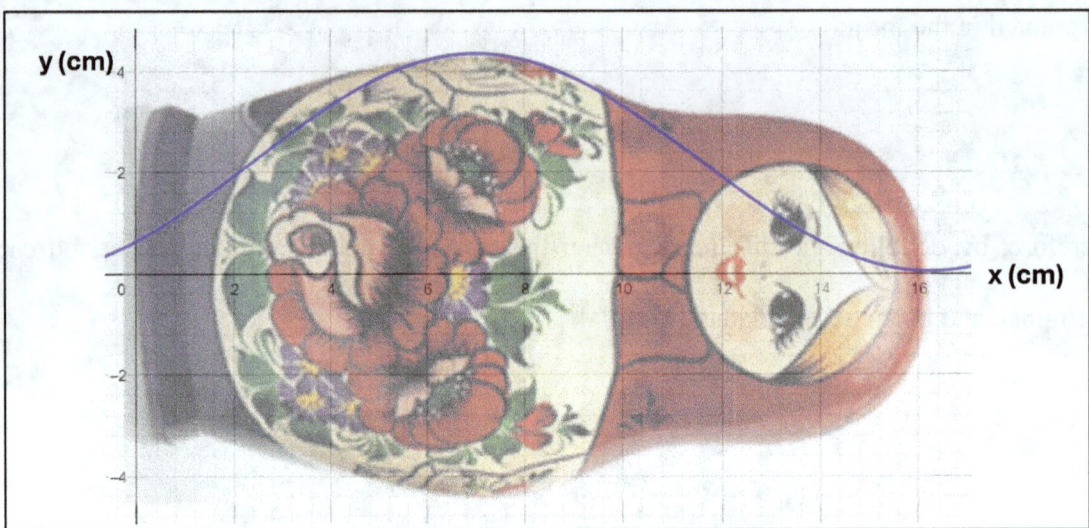

Figure 4: Graph of sinusoidal function used to model the Matryoshka doll

As seen in Figure 4, the sinusoidal function I produced and modelled does not outline the doll perfectly. It can be clearly seen that some areas of the doll are cut through and the functions does not outline the doll accurately.

D+ Good reflection

As a result of the model's poor precision, I decided to plot a different sinusoidal curve using GeoGebra. I first plotted points along the doll (as seen in Figure 5) and grouped them to create a list holding all the points. I then used the FitSin function on GeoGebra with the created list in order to produce a sine graph which had the closest fit to the points that I had selected.

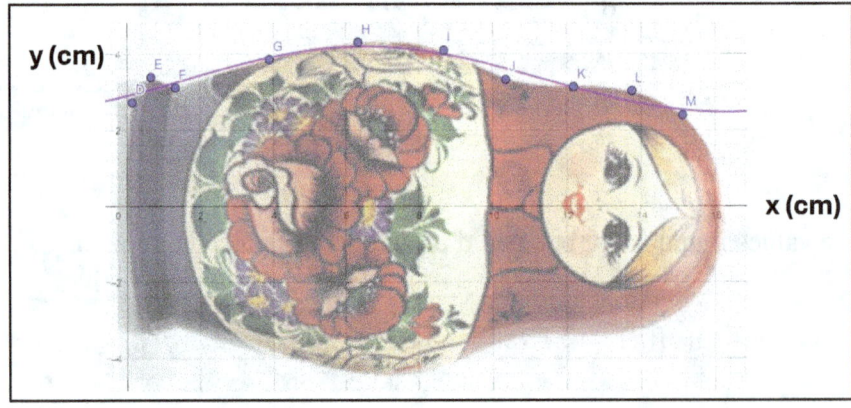

Figure 5: Graph of sine function used to model the Matryoshka doll

The equation of the sine function in Figure 5 is:

$$f(x) = 0.86 \sin(0.32x - 0.52) + 3.33$$

As seen in Figure 5, this new model is much more accurate than the model in Figure 4 as it provides a more representative outline of the true shape of the Matryoshka doll. This is because in Figure 5 the model used does not cut through large areas of the doll and produces a more elongated curve that is more indicative of the long body of the doll as opposed to the steep Function in Figure 4.

D+ again good reflection

Obviously, the function produced is still not a perfect fit for the outline of the doll as the ends are not properly modelled; the sine function doesn't curve down to fit the elliptical base and head. Additionally, the graph doesn't pass through some of the points plotted on the doll which decreases the accuracy of the model. These inaccuracies can be seen in Figure 6.

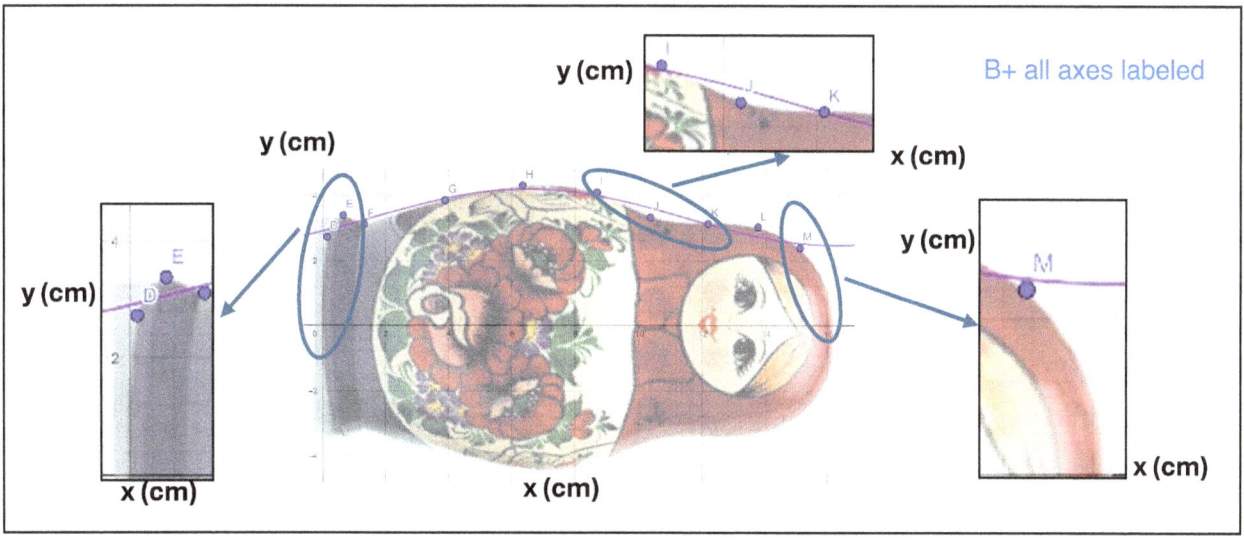

B+ all axes labeled

Figure 6: Errors within the sine model

I will still calculate the volume of the sine model as it will provide a good baseline for volume of the doll. The volume produced by the sine model will be calculated by substituting the sine function into the volume of revolution formula. The limits of the integral that I will use will be 16.2 and 0 (the real height of the doll) as I'm not concerned about capturing extra area/volume as this is only a baseline estimate. Additionally, only 2 decimal places will be used during the integral calculation for the same reason. In future models, which are more precise, the decimal places will be increased in order to utilize the increased accuracy of the model.

The volume of the sine function is calculated as follows:

E+ Mathematics is correct and good understanding demonstrated

$$V = \pi \int_0^{16.2} (0.86 \sin(0.32x - 0.52) + 3.33)^2 \, dx$$

$$V = \pi \int_0^{16.2} (0.74 \sin^2(0.32x - 0.52) + 5.72 \sin(0.32x - 0.52) + 11.09) \, dx$$

Using the double angle identity $\cos(2\theta) = 1 - 2\sin^2\theta$ and rearranging it into $\sin^2\theta = \frac{1-\cos(2\theta)}{2}$:

$$V = \pi \int_0^{16.2} \left(0.74 \left(\frac{1 - \cos(0.64x - 1.04)}{2}\right) + 5.72 \sin(0.32x - 0.52) + 11.09\right) dx$$

$$V = \pi \left[-0.58 \sin(0.64x - 1.04) - 17.88 \cos(0.32x - 0.52) + 11.09x\right]_0^{16.2}$$

$$V \approx \pi (186.46 + 15.01)$$

$$V \approx 632.95 \text{ cm}^3$$

This function was solved analytically to demonstrate the steps taken to integrate the trigonometric function but was also verified using the GDC.

D+ Critical reflection

Although I initially thought that the alternating wide and narrow segments of a sine graph would reflect the tapering of the Matryoshka doll, upon reflection, it is clear that the limited flexibility of the sine graph wouldn't properly model the varying slopes along the doll's contour. The asymmetricity of the base and top of the doll cannot be precisely modeled by the symmetrical and repetitive nature of a sine function.

Due to the inaccuracies within the sine model, I decided to use a polynomial function to model the Matryoshka doll. I thought that using a Polynomial function would be an accurate fit as a high degree polynomial would allow for multiple inflection points which would be useful for modeling the varying curvature of the doll.   D+ Critical reflection

I decided to use a polynomial to the sixth degree to model the doll as I could identify 5 turning points H, J, Q, W and A_1 (Figure 7). Although a higher degree polynomial would create a more accurate model, I decided against this to avoid overfitting the data. Overfitting occurs when the graph of the polynomial becomes too sensitive to minimal fluctuations within the data points. This would lead to curves being formed between data points and would not accurately represent the smooth, general shape of the object. Additionally, a high degree polynomial would increase the complexity the model which would make it difficult to analyze and interpret mathematically.

By using a polynomial of lower degree such as 6, a smoother and more controlled model can be created.

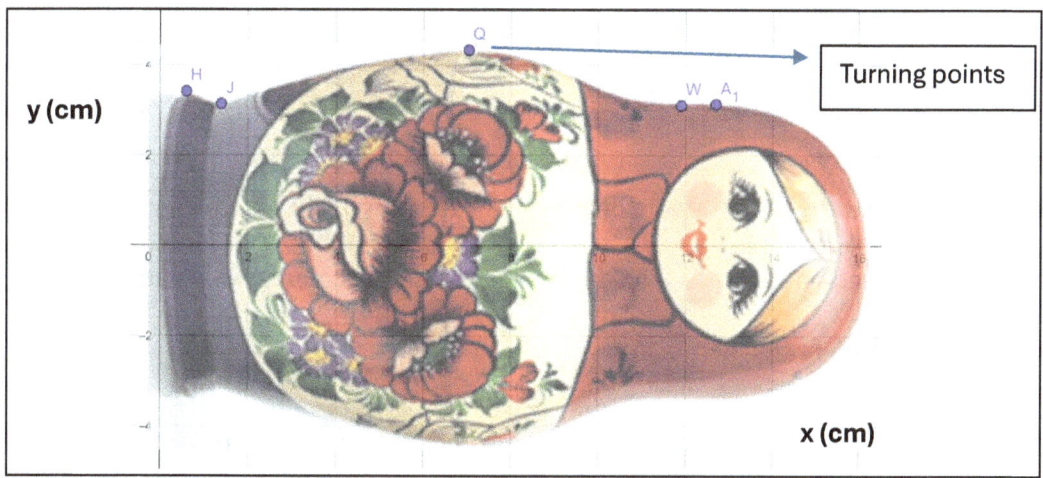

Figure 7: Identification of turning points on the Matryoshka doll

By plotting points along the outline of the doll and using GeoGebra's FitPoly tool, the following model was created:

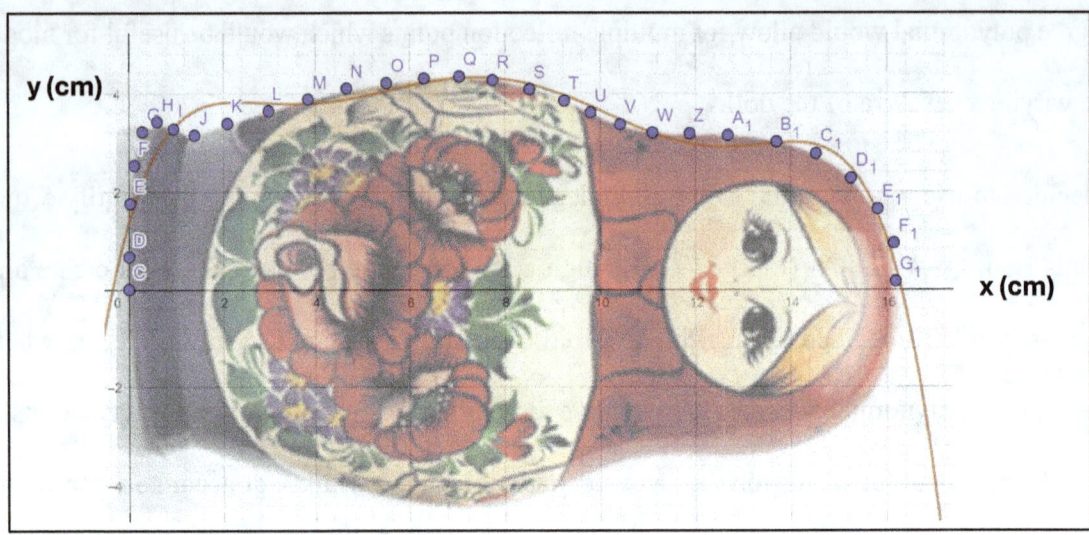

Figure 8: Polynomial function of the sixth degree to model the Matryoshka doll

The equation of the polynomial function in Figure 8 is:

$$g(x) = -0.0001x^6 + 0.0025x^5 - 0.0448x^4 + 0.388x^3 - 1.6513x^2 + 3.3001x + 1.3483$$

A four decimal place rounding was used to represent this function as it was the minimum required for the sixth degree exponent not to round to 0. However, the function in the diagram is accurate to 13 significant figures.

Rather than determining the accuracy of the model through qualitative means such as analyzing the excess area added or area cut of the doll by the graph, I decided to calculate the R^2 value, also known as the coefficient of determination.

The coefficient of determination will measure how well the modeled function passes through the selected points of the graph and as a result will provide me with a general estimation of how accurately the function models the outline of the object.

It is important to note that determining the accuracy of a model using the R^2 value may not always be reliable as the R^2 value is only a determination of how well points are passed through in a graph. This means that a graph which is curving through points on an otherwise straight line fit would show a high R^2 value but not be a good fit representing the general trend of the points. This is a problem which could arise with overfitting and is hence another reason why a low polynomial of six was selected.

D+ More good reflection

Since the polynomial function in Figure 8 is relatively smooth and not abruptly changing in direction, the R^2 value would provide valid and insightful information on the accuracy of the model outlining the Matryoshka doll.

Using the RSquare function on GeoGebra, the R^2 value was calculated to be 0.87. The 0.87 value suggests that the model is a relatively accurate fit for the outline of the doll but not perfect as a value closer to 1 would suggest.

Nevertheless, the volume will be determined for comparison between previous and future models.

Volume of the Matryoshka doll as modelled by a sixth-degree polynomial:

$$V = \pi \int_0^{16.2} (-0.0001x^6 + 0.0025x^5 - 0.0448x^4 + 0.388x^3 - 1.6513x^2 + 3.3001x + 1.3483)^2 \, dx$$

$$V \approx 631.89 \text{ cm}^3$$

This integral was solved using the GDC and no steps were shown as this is a basic integral with no trigonometric or logarithmic functions. Additionally, the function substituted into the volume of revolution formula was accurate to 13 significant figures instead of the displayed 4 decimal place rounding. Calculating the volume using a function rounded to lowest decimal places would not represent the function seen in Figure 8 and hence a misrepresentative volume would be calculated.

E+ D+ Good understanding of R squared

From Figure 9 and the calculated R^2 value of 0.87, it is clear that a polynomial model is not perfectly accurate as area above the doll is modelled and included.

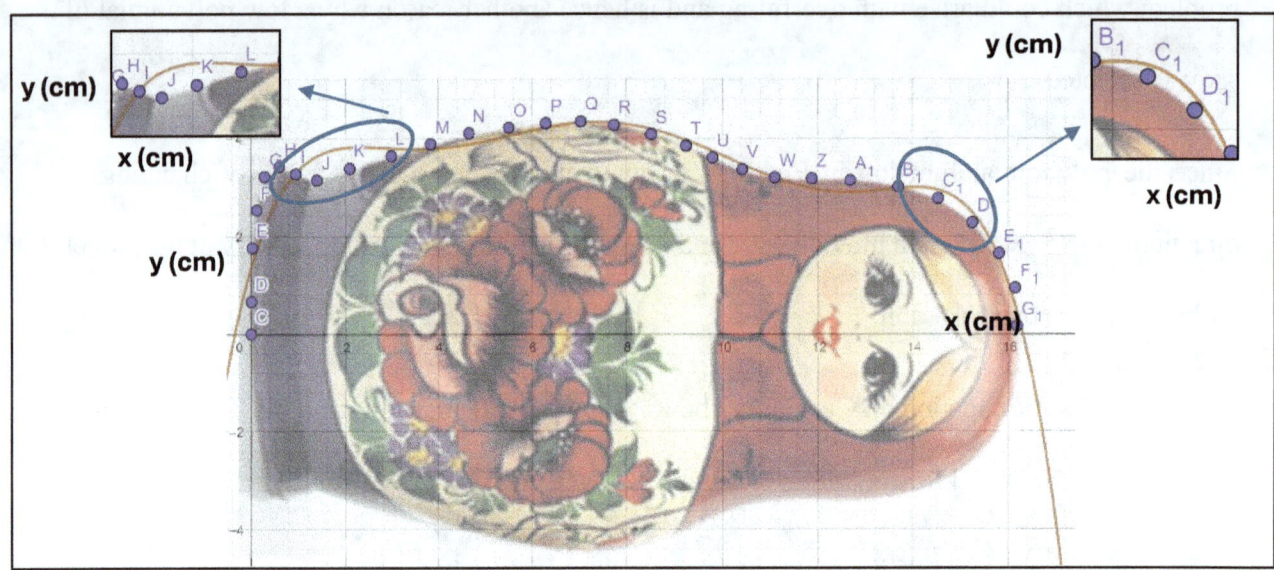

Figure 9: Errors within the polynomial model

Afterwards, I realized that if I wanted to produce an extremely accurate model of the Matryoshka doll, I would have to divide the doll into multiple sections based on its shape and model the respective outlines.

C+ Student is engaged, thinking about what they have done and what else they could do and is allowing the results to drive the exploration forward

I would achieve this by using a combination of different functions to create a piecewise function which would model the doll in its entirety.

The first function I selected was a logarithmic function as the curving base of the doll resembled the sublinear growth pattern of a log function. The graph of the log function modeling the base of the doll can be seen in Figure 10 on the next page. For convenience I will name the base of the doll Section A.

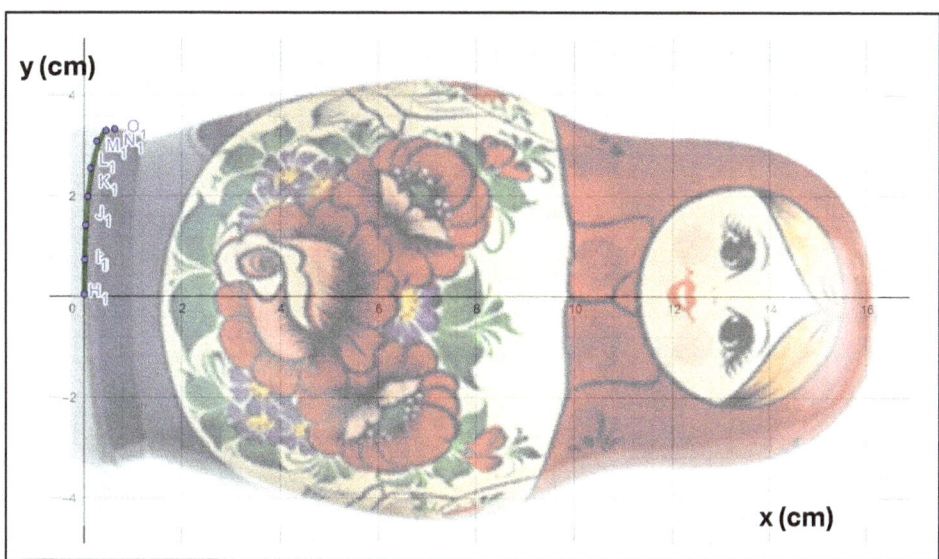

Figure 10: Logarithmic function used to model Section A

The logarithmic function model was created using the FitLog function on GeoGebra and as a results will display/produce the most accurate log function model according to the points I have plotted. The domain will range from x = 0 to x = 0.4153 as that is the range between the base of the doll and the next function intersection. *A+ The exploration is well written, clear and easy to follow*

A decimal rounding of 4 will be used throughout the modelling of the piecewise function in order to maintain the accuracy of the models and because rounding to greater number of decimal places would complicate the process as well as return an insignificant difference in the area modelled and volume calculated. *B+ significant figures mentioned and explained multiple times*

The equation of the logarithmic function including its domain in Figure 10 is:

$$f(x) = 4.0126 + 0.8485 \ln(x), \quad 0 \leq x \leq 0.4153$$

Next, a polynomial function to the second degree was used in GeoGebra to model a slight curve connecting the base of the doll to its body (Section B). I chose to use this function as I could identify one turning point and was similar in appearance to a quadratic function.

The polynomial function to model the slight curve is shown in Figure 11.

Figure 11: Polynomial function used to model Section B

The equation of the polynomial function including its domain in Figure 11 is:

$$g(x) = -0.6996x^2 + 0.9904x + 2.9764, \qquad 0.4153 \leq x \leq 1.3084$$

Next, a sinusoidal function was used to model the body of the doll (Section C) as a sine wave reflects the smooth, rounded and almost symmetrical curve of the doll's body. Using GeoGebra, the most accurate sine function was created based on the points I plotted on the doll's body (Figure 12).

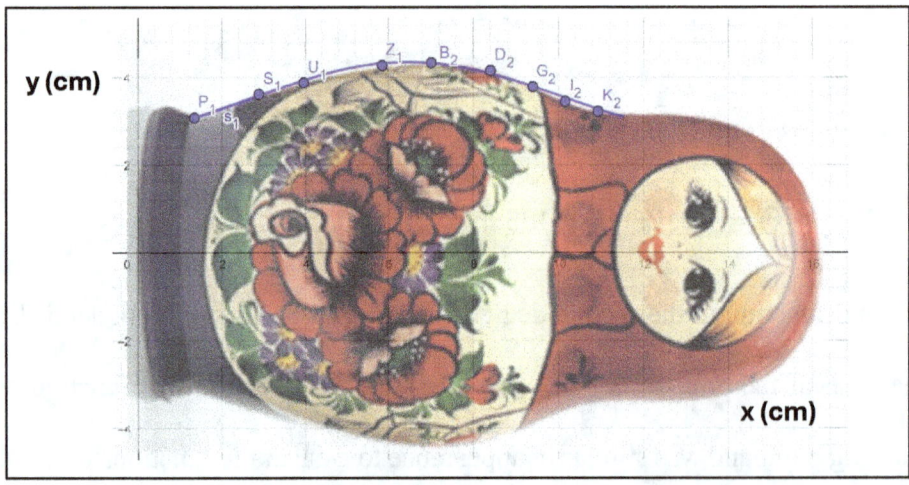

Figure 12: Sinusoidal function used to model Section C

The function of the polynomial including its domain in Figure 12 is:

$$h(x) = 3.5623 + 0.793 \sin(0.4337x - 1.2299), \quad 1.3084 \leq x \leq 11.4327$$

Finally, an ellipse was used to model the head of the doll (Section D). I initially thought to use a circle equation, however, I quickly realized that a circle equation wouldn't properly model the slightly oval shape of the head due to a circle's symmetric nature.

The ellipse tool was used to create two focus points (L_2 and M_2) in Figure 13 which when adjusted created an ellipse. This ellipse was, however, hand adjusted to match the outline of the doll's head as points couldn't be plotted along the surface of the head due to the nature of the ellipse tool. As a result, the method used to model the ellipse might be imprecise, but the resulting model remains valid based on its alignment with the doll's head.

The ellipse used to model the head of the doll is shown in Figure 13.

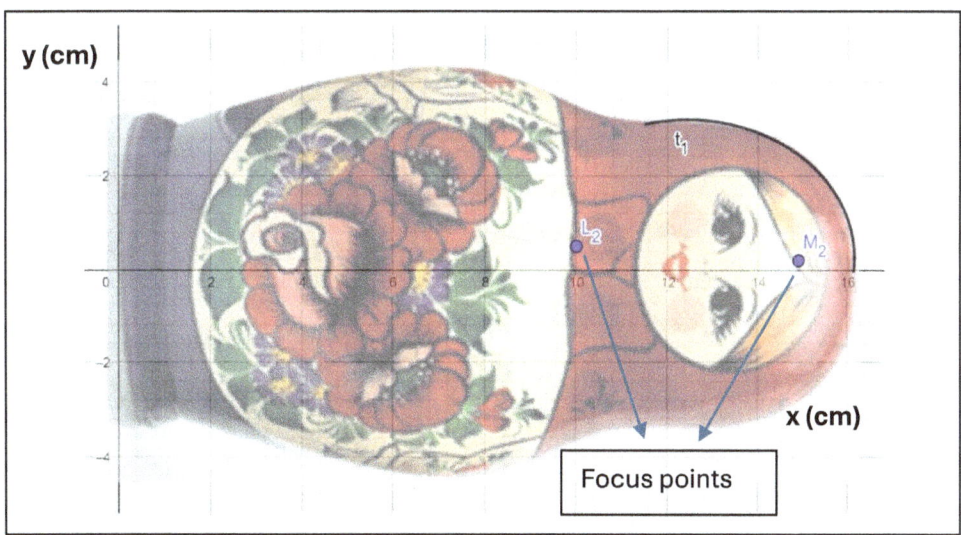

Figure 13: Ellipse used to model Section D

The equation of the ellipse is written in the form $\frac{(x-h)^2}{a^2} + \frac{(y-k)^2}{b^2} = 1$ where (h, k) are the coordinates of the center of the ellipse, a is the horizontal radius of the ellipse and b is the vertical radius of the ellipse (Cuemath 2022).

The equation of the ellipse in Figure 13 including its domain and range is:

$$\frac{(x-12.44)^2}{3.71^2} + \frac{(y-0)^2}{3.21^2} = 1, \quad 11.4327 \leq x \leq 16.2, \quad y \geq 0$$

Transforming this equation into a function gives:

B+ E+. Well communicated and good understanding of mathematics.

$$i(x) = \sqrt{3.21^2 \left(1 - \frac{(x-12.44)^2}{3.71^2}\right)}, \quad 11.4327 \leq x \leq 16.2$$

Combining the equations together produces the piecewise function below and graph in Figure 14.

$$p(x) = \begin{cases} 4.0126 + 0.8485 \ln(x) & 0 \leq x \leq 0.4153 \\ -0.6992x^2 + 0.9904x + 2.9764 & 0.4153 \leq x \leq 1.3084 \\ 3.5623 + 0.793 \sin(0.4337x - 1.2299) & 1.3084 \leq x \leq 11.4327 \\ \sqrt{3.21^2 \left(1 - \frac{(x-12.44)^2}{3.71^2}\right)} & 11.4327 \leq x \leq 16.2 \end{cases}$$

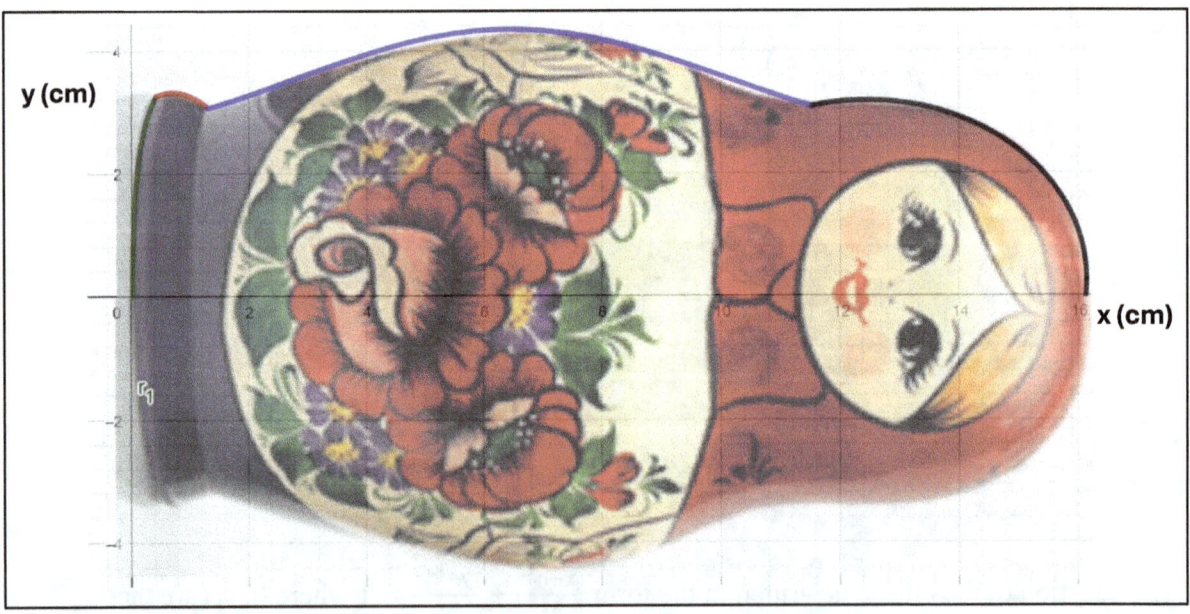

Figure 14: Piecewise function used to model the Matryoshka doll

C+ This is nicely created unique piecewise model of the doll

Using GeoGebra 3D, I produced a 3D render of the piecewise function I used to model the matryoshka doll as seen in Figure 15.

C+ Nice use of software to create image

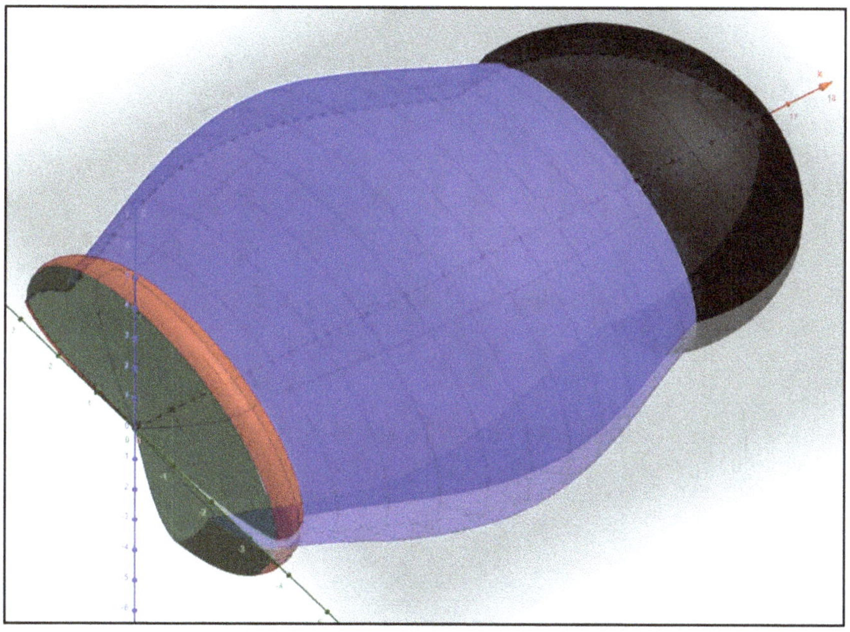

Figure 15: 3D render of the piecewise function

To determine the volume modeled by the piecewise function, the volumes of the individual sections must be calculated and added together. This was done by substituting the functions into the volume of revolution formula and verifying the answer with the GDC.

Section A:

$$V_1 = \pi \int_0^{0.4153} (4.0126 + 0.8485 \ln(x))^2 \, dx$$

$$V_1 = \pi \int_0^{0.4153} (16.1009 + 6.8094 \ln(x) + 0.72 \ln(x)^2) \, dx$$

B- ln(x) should be in a bracket here although the maths follows correctly as if it is

Splitting the integral into three terms:

$$V_1 = \pi \left(\int_0^{0.4153} 16.1009 \, dx + \int_0^{0.4153} 6.8094 \ln(x) \, dx + \int_0^{0.4153} 0.72 \ln(x)^2 \, dx \right)$$

First Term:

$$\int_0^{0.4153} 16.1009\, dx = [16.1009x]_0^{0.4153}$$

Second Term:

Using integration by parts:

$$\int \ln(x)\, dx = x\ln(x) - x + C$$

$$\int_0^{0.4153} 6.8094 \ln(x)\, dx = [6.8094(x\ln(x) - x)]_0^{0.4153}$$

Third Term:

Using integration by parts again:

$$\int u\, dv = uv - \int v\, du \qquad \text{Let } u = \ln(x)^2$$

$$du = 2\ln(x) \times \frac{1}{x} dx = \frac{2\ln(x)}{x} dx$$

Let $dv = dx$, $v = x$

E+ Rigor and sophistication demonstrated

Therefore:

$$\int \ln(x)^2 dx = x\ln(x)^2 - \int x \times \frac{2\ln(x)}{x} dx$$

$$\int \ln(x)^2 dx = x(\ln(x)^2 - 2\ln(x) + 2)$$

$$\int_0^{0.4153} 0.72\ln(x)^2 dx = [0.72(x(\ln(x)^2 - 2\ln(x) + 2))]_0^{0.4153}$$

Summing the three terms together:

$$V_1 = \pi\left[16.1009x + 6.8094(x\ln(x) - x) + 0.72(x(\ln(x)^2 - 2\ln(x) + 2))\right]_0^{0.4153}$$

$$V_1 = \pi(2.7282 - 0)$$

$$V_1 \approx 8.5709 \text{ cm}^3$$

Section B (Steps not shown as the integral is simple):

$$V_2 = \pi \int_{0.4153}^{1.3084} (-0.6992x^2 + 0.9904x + 2.9764)^2 \, dx$$

$$V_2 \approx 29.9078 \text{ cm}^3$$

Section C:

$$V_3 = \pi \int_{1.3084}^{11.4327} (3.5623 + 0.793\sin(0.4337x - 1.2299))^2 \, dx$$

$$V_3 = \pi \int_{1.3084}^{11.4327} (0.6288\sin^2(0.4337x - 1.2299) + 5.6526\sin(0.4337x - 1.2299) + 12.692) \, dx$$

Using the double angle identity $\cos(2\theta) = 1 - 2\sin^2\theta$ and rearranging it into $\sin^2\theta = \frac{1-\cos(2\theta)}{2}$:

$$V_3 = \pi \int_{1.3084}^{11.4327} \left(0.6288 \times \frac{1 - \cos(0.8674x - 2.3498)}{2} + 5.6526\sin(0.4337x - 1.2299) + 12.692\right) dx$$

$$V_3 = \pi\left[13.0064x - 0.3625\sin(0.8674x - 2.3498) - 13.0334\cos(0.4337x - 1.2299)\right]_{1.3084}^{11.4327}$$

$$V_3 = \pi(159.2032 - 7.0797) \approx 477.9101 \text{ cm}^3$$

Section D: E+ Mathematics is good. Perhaps some words between equations would have helped the reader follow.

$$V_4 = \pi \int_{11.4327}^{16.2} \left(\sqrt{3.21^2 \left(1 - \frac{(x-12.44)^2}{3.71^2}\right)} \right)^2 dx$$

$$V_4 = \pi \int_{11.4327}^{16.2} \left(3.21^2 - \frac{3.21^2(x-12.44)^2}{3.71^2} \right) dx$$

$$V_4 = \pi \int_{11.4327}^{16.2} 0.7486(-x^2 + 24.88x - 154.7536) + 10.3041$$

$$V_4 = \pi [-0.2495x^3 + 9.3129x^2 - 105.5478x]_{11.4327}^{16.2}$$

$$V_4 = \pi(-326.553 + 362.274)$$

$$V_4 \approx 112.221 \text{ cm}^3$$

Summing up all the volumes of the separate sections gives the overall volume of the piecewise function as calculated below:

$$V_1 + V_2 + V_3 + V_4 = V_{Total}$$

$$8.5709 + 29.9078 + 477.9101 + 112.221 = 628.61 \text{ cm}^3$$

The 628.61 cm³ volume of the piecewise function is slightly less than the 631.89 cm³ volume of the polynomial function and the 632.95 cm³ volume of the sinusoidal function. This suggests that the piecewise function is a more accurate model as it excludes the excess area identified in the other functions.

Since the piecewise function is a product of multiple functions, individual R^2 values for each function must be determined to assess the over accuracy/fit of the piecewise function.

The individual R^2 value for each function is below.

Logarithmic function: $R^2 = 0.9637$

Polynomial function: $R^2 = 0.9392$

Sinusoidal function: $R^2 = 0.9795$

Elliptical function: $R^2 = $ N/A

As mentioned before, the ellipse was created using the ellipse tool in GeoGebra and therefore no points were plotted along the outline of the doll's head. As a result, an R^2 value can't be determined for the elliptical function. However, since three of the four functions have an R^2 value greater than 0.93 with one being close to 0.98, it can be safe to conclude that the overall piecewise function is highly accurate fit to the plotted points.

E+ Again good understanding of R squared and explanation as to why it is not used for the ellipse

Convergence volume

Now that the most precise volume of the largest Matryoshka doll has been calculated, by relating the width and heights of the largest and following Matryoshka doll, a scale factor can be determined and the volume of the smaller dolls can be calculated.

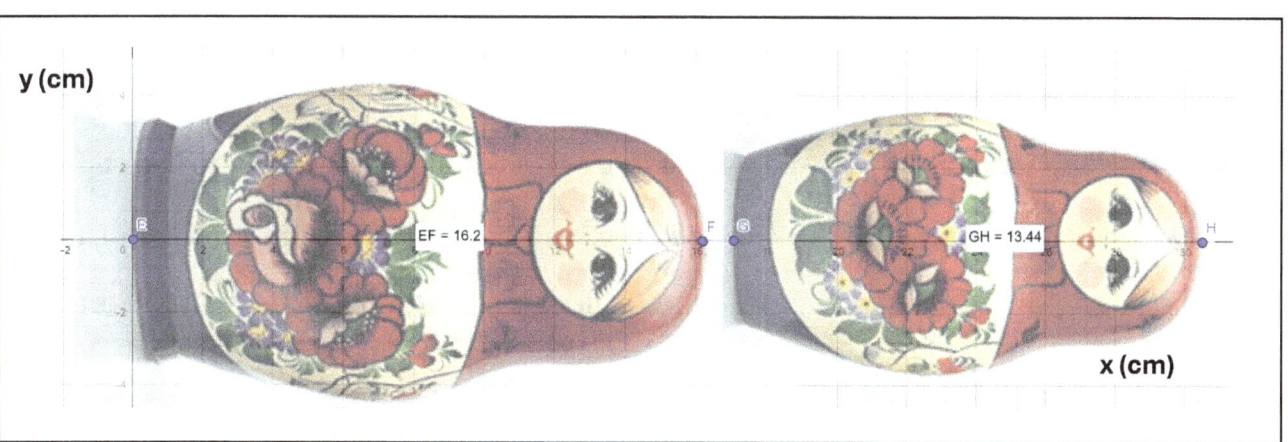

Figure 16: Heights of the largest and following Matryoshka doll in the same set

In Figure 16, the volume of the following Matryoshka doll can be calculated by:

$$V_2 = k^3 \times V_1$$

Since the scale factor (k) is $\frac{13.44}{16.2} = 0.8296$, and $V_1 = 628.61$, $V_2 = 0.8296^3 \times 628.61$.

$$V_2 \approx 358.912 \text{ cm}^3$$

Assuming that consecutive dolls share the same scale factor, by using the finite geometric series formula as seen below, the convergence volume of the dolls following the largest doll (S_n) can be calculated and then compared.

A+ E+ C+ Really nice addition to the IA to look at the scale fac and then sum as a sequence. This was a unique idea and rele* to the aim*

$$S_n = a \times \frac{1 - r^n}{1 - r} \qquad (\text{for } r \neq 1)$$

There are 7 dolls following the largest doll in my grandmothers set so n (the number of terms) will equal 7, r (the common ratio) will equal $\frac{358.912}{628.61}$ (0.571) and a (the first term) will equal 358.912.

Substituting the values into the finite geometric series formula gives:

$$S_n = 358.912 \times \frac{1 - 0.571^7}{1 - 0.571}$$

$$S_n = 820.068 \text{ cm}^3$$

Using water displacement, I measured the actual volumes of the 8 Matryoshka dolls in the set. I found out that the largest one had a volume of 629 cm³ and the sum of the following dolls had a volume of 820 cm³. These values are very similar to the volumes of the Matryoshka dolls that I obtained, suggesting that my measurements, scaling and models were all accurate to a high degree.

Conclusion/Evaluation

A+ nice conclusion which ties back to the aim.

The aim of this investigation was to find the best model to calculate the volume of the largest Matryoshka doll and by using a scale factor, determine the cumulative volume of the following dolls within. This aim was carried out successfully as three different models were used to determine the volume of the largest doll where each model become progressively more precise and accurate as illustrated by the decreasing volumes of Model 1: 632.95 cm^3, Model 2: 631.89cm^3 and Model 3: 628.61 cm^3. This allowed me to obtain the most accurate convergence volume of the proceeding dolls which was calculated to be 820.068 cm^3.

Although the final model used to model the largest Matryoshka doll might not have been perfect, with minor inclusion of excess areas observed, the close approximation of less than 1 cm^3 to the real-life volume of the largest doll suggests that the modeled errors were minimal and that the model used is still highly accurate. Additionally, the assumption that subsequent dolls possessed identical scale factors could have potentially hindered the validity of the exploration if a convergence volume similar to the real-life convergence volume of the dolls was not found. Fortunately, this was not the case as yet again a margin of error of less than 1 cm^3 to the real life convergence volume was found, suggesting that the assumption used was accurate and valid.

The rounding to four decimal places within the final model also ensured a high level of precision and accuracy which contributed to the validity of the exploration.

This exploration could also be extended to determine the surface area of each Matryoshka doll in order to estimate the amount of paint needed for each doll. This would provide information on how paint cost would scale with decreasing sizes and provide insight into the material costs for manufacturing and designing these painted dolls.

Bibliography

Cuemath (2022). *Ellipse - Equation, Properties, Examples | Ellipse Formula*. [online] Cuemath. Available at: https://www.cuemath.com/geometry/ellipse/.

IA Sample 8 Examiner Comments: Guitar Area

This is an excellent high-achieving HL IA where the student uses modelling to find the area of their guitar.

Author: Armaan Hussein

Appropriate for: AAHL, AIHL, AASL, AISL (A simpler version for SL)

Criterion	SL	HL
A	3	3
B	4	4
C	4	4
D	3	3
E	6	6
Total	19	19

A: Exploration is coherent and well organised. It is well written and easy to follow. There is a good flow from one section to the next. The only reason A4 was not awarded is because the exploration is a bit too long and it perhaps loses some conciseness.

B: The mathematical communication is excellent. There is so much detail put into every graph. This is marked on the paper. Mathematical notation and terminology are correct. Axes are labelled and all variables defined. Rounding is explained and consistent. Struggled to find any errors. B4 was easily awarded.

C: Engagement is outstanding. Student is clearly fully invested in the exploration and the mathematics. They have a strong interest in their work and the engagement drives the exploration. Multiple examples are marked on the paper. C4 awarded.

D: Many examples of critical reflection noted on the paper. The student regularly stops, looks at their results and reflects. The reflection helps them choose the path forward. Multiple examples are marked on the paper, e.g. page 4, 5, and 6. So much good quality and critical reflection along with complete understanding led to D3.

E: The mathematics is commensurate with the level of the course. The understanding is thorough. The student's explanations are thorough, and they demonstrate a complete understanding of what they are doing. Sophistication and rigour are demonstrated. Integration by parts on page 19 is an example of this.

1.0 Introduction

The legendary Eddie Van Halen once said, "A guitar is a very personal extension of the person playing it."

Having had a guitar in my hand since age 5, it has become an inextricable part of my life. As a guitarist, taking care of my instrument is incredibly important because of my sentimental connection from using it in every concert I have played since I was gifted it for my 13th birthday. To protect my guitar, I apply Brazilian Carnauba wax monthly to its body, mainly on the soundboard of the guitar (refer to Figure 1.1.1), to protect it from moisture, sweat, and dust, which comes from frequent use. Since it is an integral part of my guitar maintenance process, I was intrigued to figure out how much guitar wax I need for a year.

Figure 1.1.1
Annotated Guitar Image with Dimensions

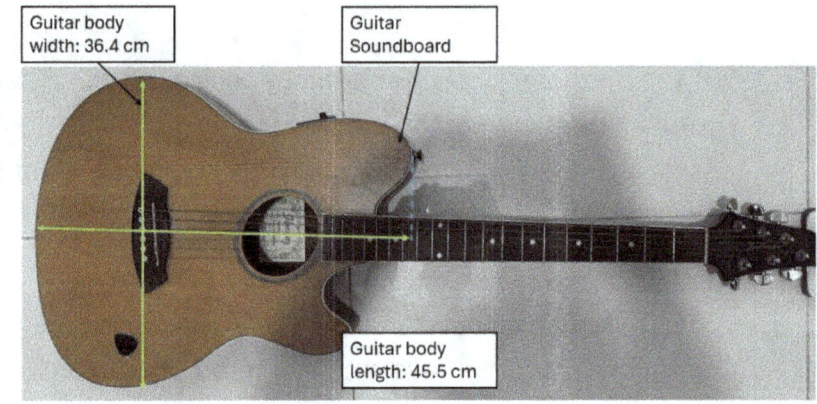

A+ Clear aim

Therefore, the aim of this exploration is to calculate the area of the guitar soundboard to calculate the volume of guitar wax I need for a year.

First, I measured the length and width of my guitar's soundboard, as seen in Figure 1.1.1. I then uploaded a photo of my guitar's body to GeoGebra. This online mathematics software allows me to size my photo to the correct dimensions and plot, manipulate, and transform functions to model the shape of my guitar's soundboard. I then tried various functions to model the different curves of my guitar, such as sine functions, polynomial functions, and logarithmic functions, amongst others. Different functions fit different curves the best, and a combination of multiple functions was used to model the shape of the soundboard. The area of the sound hole was calculated by measuring its dimensions and applying them to the ellipse area formula.

Then, the calculated area found from integrating the functions was subtracted by the sound hole area and guitar neck area, which protruded into the soundboard. Varying online sources give a range of values of the wax layer's thickness, such as 40 nm, from the acoustic guitar forum. Considering that the applied wax is not perfectly applied, and the cloth used to apply also absorbs some, an assumption of 0.2 mm was taken as the thickness of a wax layer, equivalent to the thickness of the final coating of a guitar in production (Yamaha, n.d.). A+ Very well written and easy to follow

The Guitar body was split into four sections.

Figure 1.2.1

Diagram Showing Hand-drawn Curves of the Guitar Body

C+ nicely created image of the guitar with sections

As seen in Figure 1.2.1, There are two sections to model the upper and lower curves of the guitar. Shown in Figure 1.2.2, I chose to separate the modelling of the guitar horns so that I could use functions of x to model curves two and four by changing the orientation of the guitar to be vertical.

Figure 1.2.2

Diagram Showing Hand-drawn Guitar Horn Curves

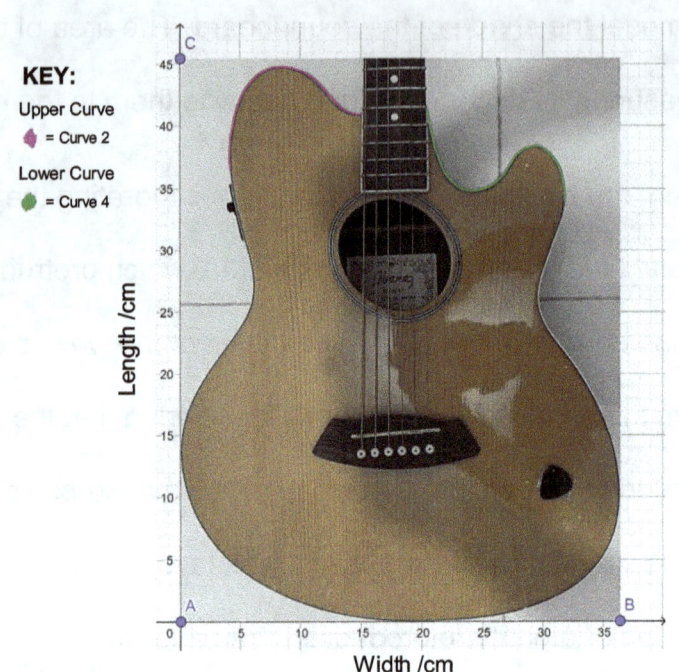

2.1 Modelling the First Guitar Body Upper Curve

Using Cubic Functions

As seen in Figure 2.1.1, I recognized that the shape of the guitar rises to a turning point and then lowers down to another turning point before continuing to rise back up. Therefore, with the two turning points, I plotted a cubic function to model the shape of the curve. However, considering the large scale of the guitar body on the cartesian plane, it is difficult to transform a cubic function accurately; therefore, I used the GeoGebra fit function to create a cubic function by plotting points along the upper curve.

C+ Good use of engagement to drive the exploration. Student is thinking about the shape of the guitar before choosing which functions to use.

When using equations generated by the GeoGebra fit function, rounding the values to 3 significant figures (s.f) causes the function to deviate from the generated function.

However, when rounded to 5 s.f, they minimally alter the fit line graphs and maintain accuracy.

Figure 2.1.1

Graph showing accuracy difference between graphed equations rounded to 3 and 5 significant figures

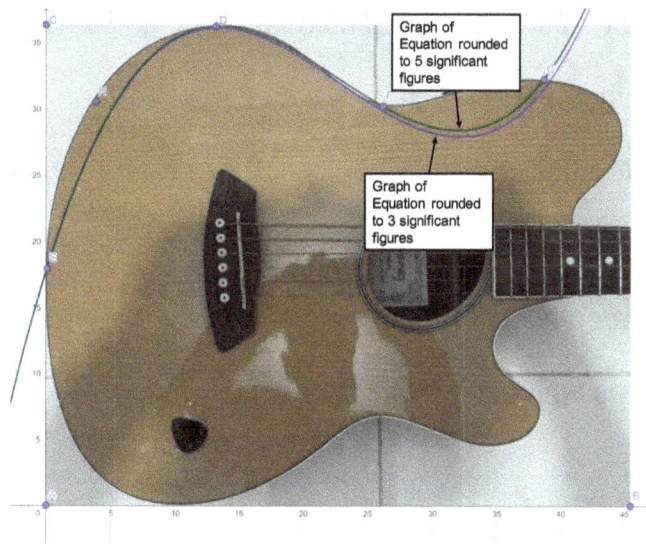

Hence, values were rounded to 5 s.f for equations generated using the fit function.

Below is the graph of a cubic equation generated using GeoGebra's fit function.

Figure 2.1.2

Graph Modelling Guitar Upper body with a Cubic Function

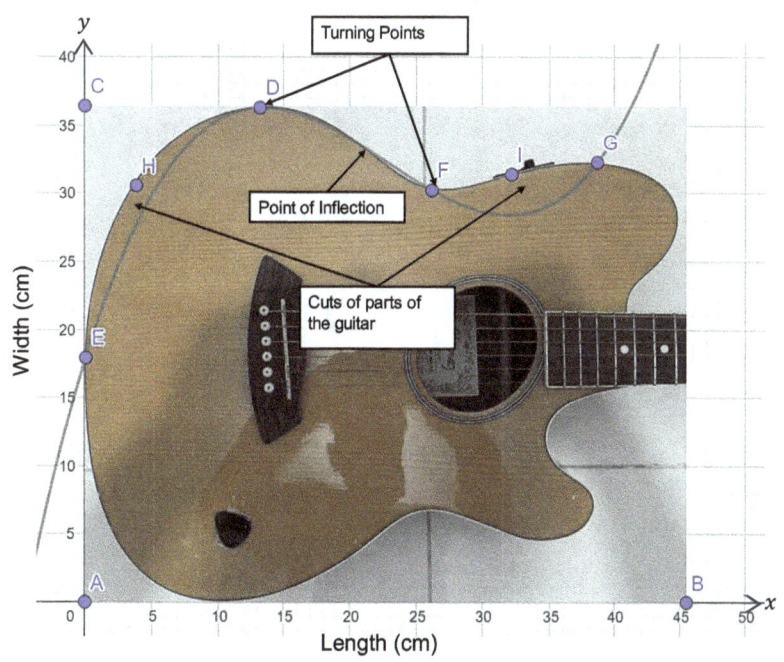

The graph of the function in Figure 2.1.2 is as follows:

$$a(x) = 0.00249x^3 - 0.16961x^2 + 3.21347x + 17.69953$$

Looking at Figure 2.1.2, the graph of the cubic function cuts off significant parts of the guitar body. It does not effectively accommodate the guitar body's upper curve because the rate of change of the function makes it steeper than the curve. Hence, I deduced that one single cubic function could not model the upper curve by itself. Because I am considering the first upper curve as a cubic equation, it means there is a point of inflection between its turning points. So, I decided to model the upper curve using two cubic equations which model the guitar curve before and after the point of inflection separately.

C+ D+ Again, engagement and reflection is guiding the exploration a helping the student decide which models to choose

Figure 2.1.3

Graph Re-modelling Guitar Upper body with Cubic Functions

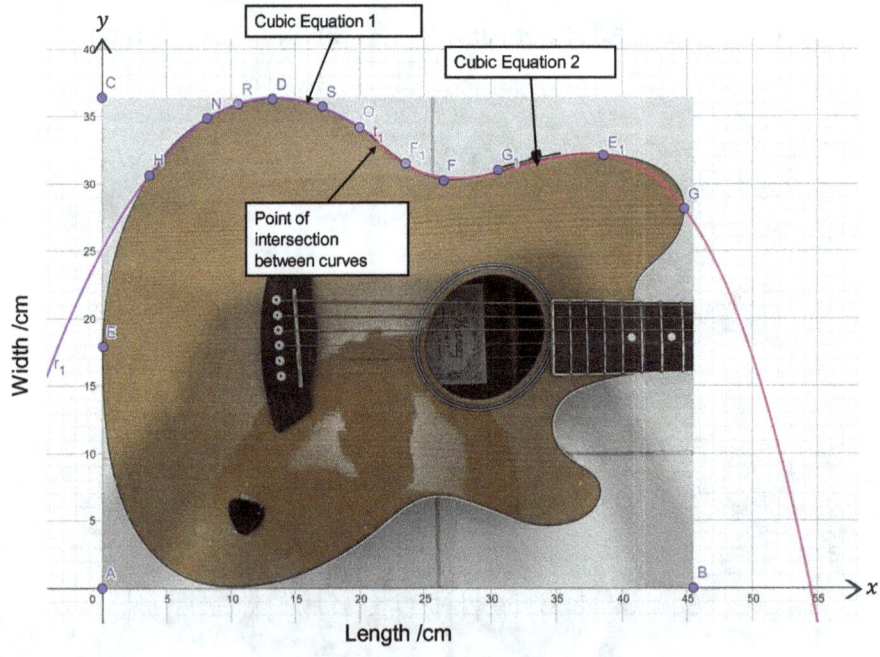

Cubic Equation 1:

$$b(x) = (7.1564 \times 10^{-4})x^3 - 0.082640x^2 + 1.8245x + 12.047, x < 21.056$$

Cubic Equation 2:

$$c(x) = -0.0035146x^3 + 0.34278x^2 - 10.873x + 143.29, x \geq 21.056$$

The two cubic functions of Figure 2.1.3 fit the shape of the guitar well and have R^2 values of 0.99913 and 0.98892 (5 s.f). These values are close to 1 because GeoGebra uses regressions to generate a function that best fits the points placed along the guitar, but the function does not necessarily fit the guitar perfectly.

There are two areas where the graph cuts slightly into the guitar body; similarly, the second cubic function misses a slight part of it. The difference between the area cut and overcompensated for might make the inaccuracy small enough not to impact the calculation of the soundboard's area significantly.

Figure 2.1.4 *Very good reflection*

Annotated Image Showing Fit of Cubic Functions

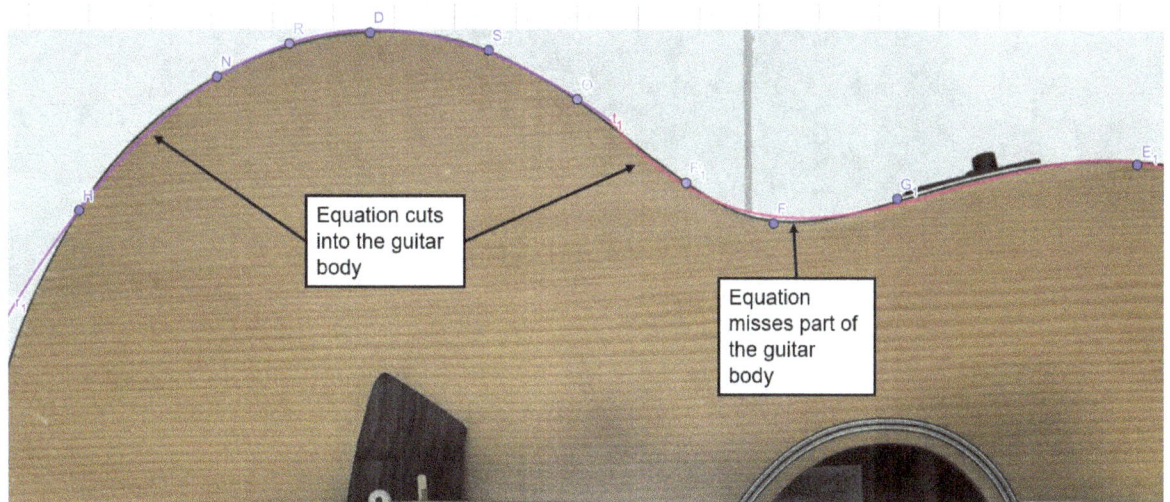

However, I chose to investigate if there was a function that could better fit the curve.

Using Sine Functions

The general equation for a sine function is:

$$y = A \times \sin(Bx + C)$$

Where A is the initial amplitude or height of the function; B determines the horizontal stretch and C horizontally translates the function (Mathematics Libre Texts, n.d.).

Like a sine curve, the guitar shape has maximum points or peaks. Considering the first upper curve as a sine curve, its peaks are unaligned, unlike a sine function.

Figure 2.1.5

Annotated Image Showing the Upper Curve's Unaligned Peaks

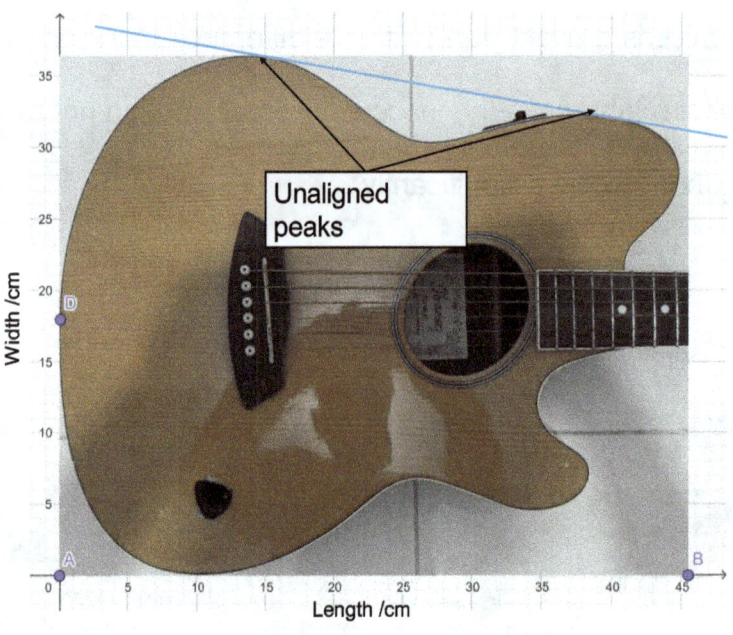

A+ Again well explained. Exploration is easy to follow.

However, if I rotate the image of the guitar these peaks align and make it possible to model the upper curve using a sine curve.

Figure 2.1.6

Annotated Image Showing Rotated Image of Guitar Body

Instead of rotating the image of the guitar body, I discerned that I could transform a sine function to ascend or descend to fit the upper curve. Hence, I added the sine function to a linear function represented by the value $Dx + E$ where I could change the gradient value represented by D, to manipulate the sine function to ascend or descend, and the value E to vertically translate the function. Therefore, the general equation of the sine and linear function is as follows:

E+ This is sophisticated mathematics and well beyond what

$$y = A \times \sin(Bx + C) + Dx + E$$

First, I plotted a combined sine and linear function (in Figure 2.1.7 below), where I set the linear function value of E to 7 so that the function is translated onto the upper curve of the guitar body. I manipulated the function's shape by using an initial value of 16.3 for A, as it is close to half the width of the guitar and reduces the amplitude significantly so that one wavelength of the sine function is visible. I used 0.15 for the x-coefficient within the sine part of the function so that one period of the function's local maxima and minima align with the turning points and a gradient value of 1 as the second turning point of the upper curve is much higher than the initial principal axis of the first sinusoidal.

D+ Very good reflection

Figure 2.1.7

Annotated Graph of Sine Function on the Guitar Body

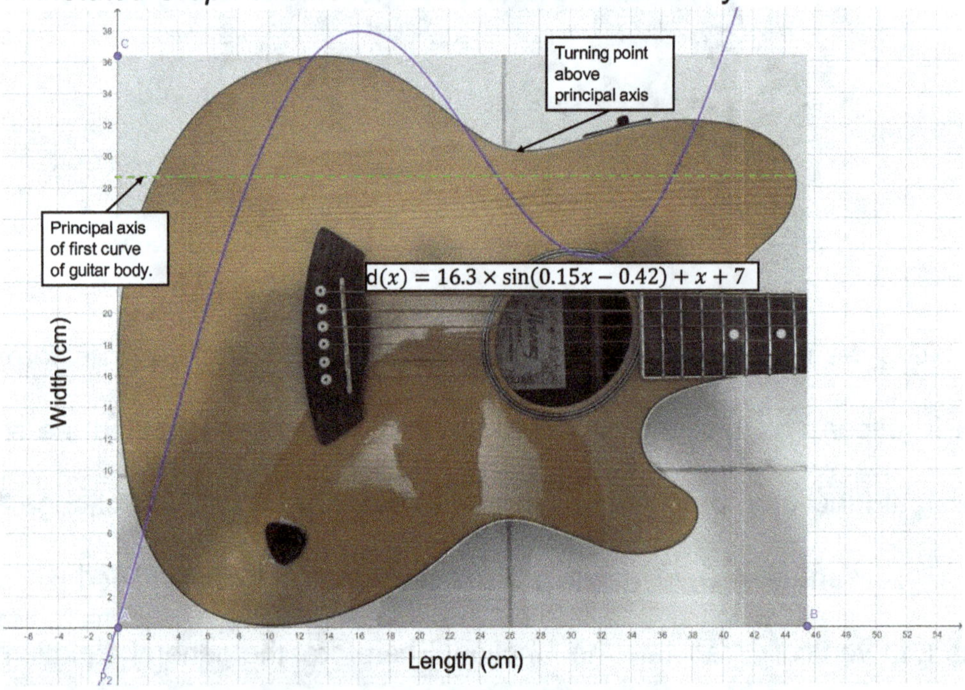

$$d(x) = 16.3 \times \sin(0.15x - 0.42) + x + 7$$

I noticed that the function did not fit well and had to be horizontally stretched by a greater scale factor; the second turning point needed to be positioned higher, and the amplitude had to be smaller. I made the relevant changes to produce the adjusted graph below.

Figure 2.1.8

Annotated Graph of Adjusted Damped Sine Function on the Guitar Body

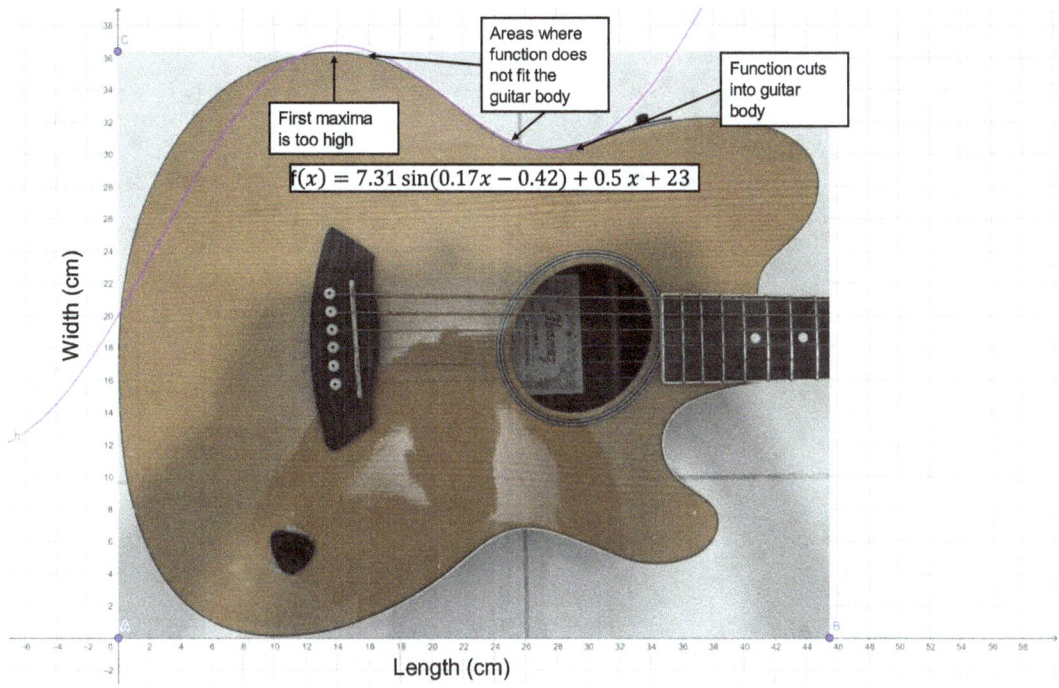

In Figure 2.1.8, to stretch the function horizontally by a greater scale factor, I increased the coefficient of x within the sine function from 0.17 to 0.15. I changed the initial amplitude to 7.31 to reduce the amplitude and decreased the linear gradient to 0.5 so that the point of inflection lies on the guitar's body. Conversely, compared to the cubic functions, the area under the function that does not fit the guitar body is much greater than the area cut into the guitar. Hence, the cubic functions better model the curve of the guitar body.

On the other hand, neither of the functions models the initial rising curve of the upper curve. Hence, I explored how to model that curve with other functions.

D+ C+ All good reflection and personal engagement driving the exploration

Using Logarithmic Functions

Figure 2.1.9

Graph of Combined Cubic Functions on Guitar Body

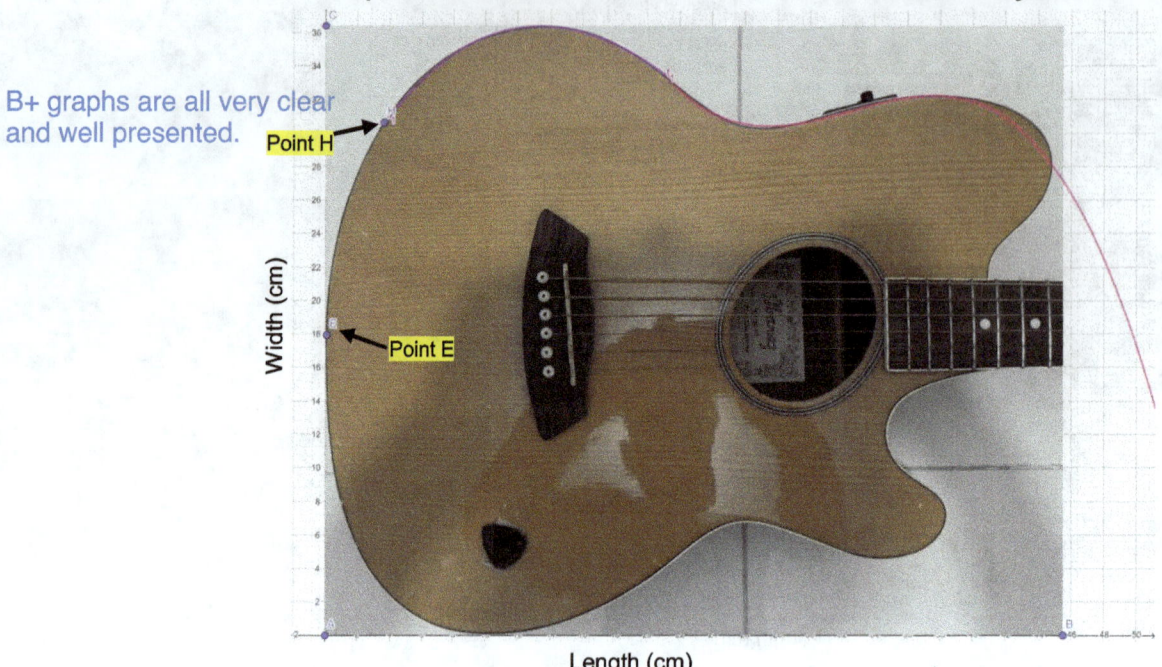

B+ graphs are all very clear and well presented.

Length (cm)

Looking at Figure 2.1.9, a function that rises at a decreasing rate of change would be required to model the curve of the guitar body between points E and H. Therefore, I decided to transform and translate a logarithmic function.

According to Lumen Learning and OntarioTech University, a logarithmic function can be transformed and translated by changing the following values:

$$y = A\log(Bx + C) + D$$

A - Controls the vertical stretch where values greater than one function are stretched and made steeper.

B - Controls the horizontal stretch scale factor.

C - Horizontally translate the function.

D - Vertically translates the function.

Recognizing that the logarithmic function would have to be very steep, the function would need values greater than 1 for A and B. However, the function's domain would be very small as $x = 3.65$. This small domain means that a much larger value for B is needed compared to A to horizontally stretch the function into the range. I chose to transform a natural logarithmic (ln) function as it will simplify integrating the soundboard's area.

Hence, I first attempted modelling the initial curve using an ln function, as seen in the graph below.

Figure 2.1.10

Graph of Initial Natural Logarithmic Function on Guitar Body

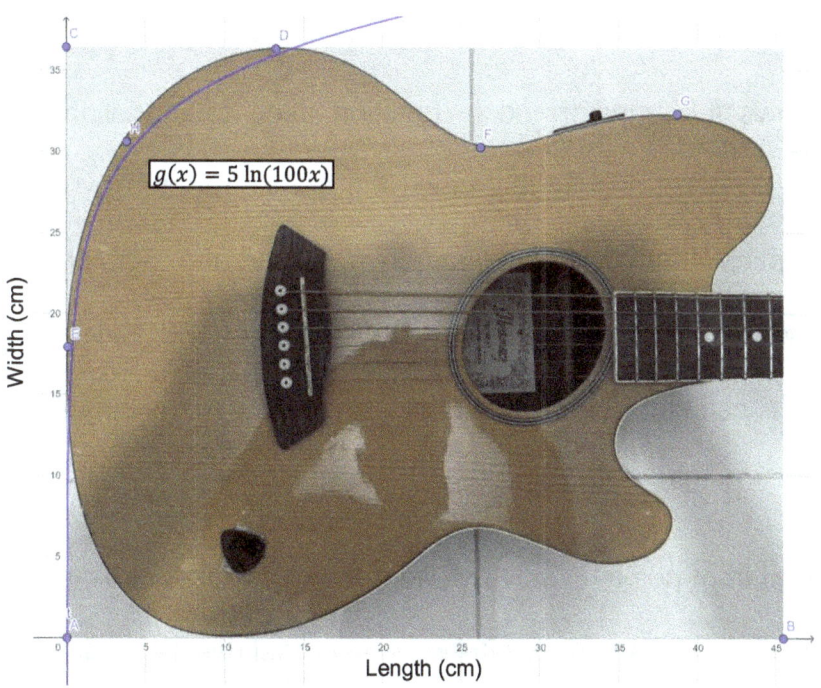

Figure 2.1.10 shows the graph of the function $g(x) = 5 \ln(100x)$. I discerned that it does not fit the guitar's curve between points E and H well. However, I recognize that the function can be translated so that it passes through point E and slightly stretched vertically so that it passes through point H as well.

Figure 2.1.11

Graph of Transformed Natural Logarithmic Function

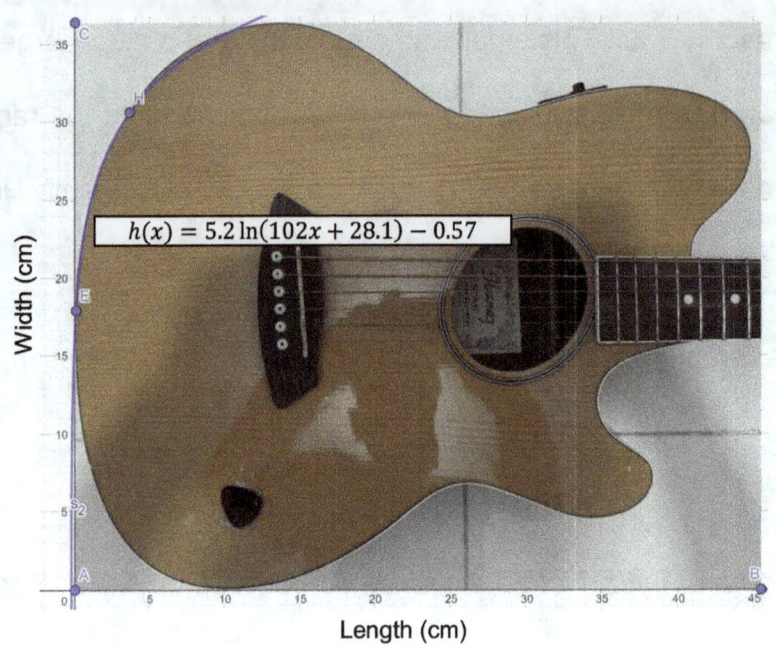

Figure 2.1.11 shows the transformed ln function used to model the initial curve between points E and H. I slightly increased the vertical stretch by multiplying the function by 5.2 rather than 5. In addition, I increased the horizontal stretch by multiplying the x value within the function by 102 instead of 100. Since the function is horizontally compressed significantly, the horizontal translation is larger; however, the vertical translation is relatively smaller.

It fits the guitar well from points E to H, where it meets the first cubic equation of the upper curve. Therefore, I will use this ln function to model this part of the guitar body's curve.

A+ D+ Good explanations and reflections to arrive at a nice function

Equation of the ln function:

$$h(x) = 5.2 \ln(102x + 28.1) - 0.57$$

2.2 Modelling the Second Guitar Body Upper Curve

For the second curve of the guitar's upper body, I similarly recognized the applicability of cubic and the combined sine and linear functions. Hence, I carried out a regression process utilising Geogebra's fit function to generate a combined sine and linear function as well as a cubic function to fit the guitar horn's shape.

Figure 2.2.1

Graph of the Second Upper Curve

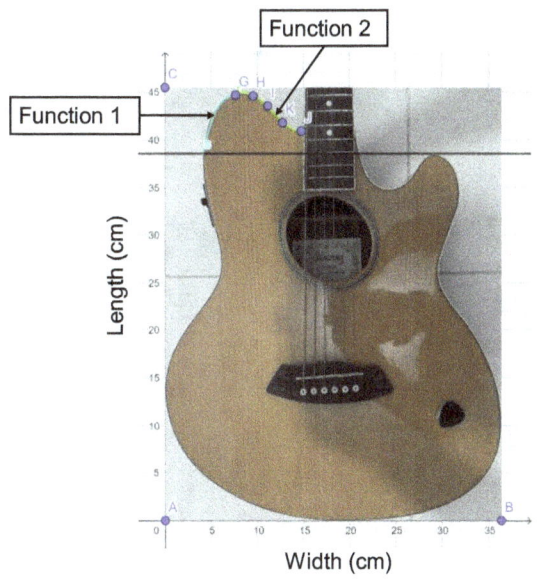

Upper guitar horn function 1:

$$j(x) = -9482.75 \sin(0.041738x + 5.9609) + 395.97x - 3012.8$$

Upper guitar horn function 2:

Less explanations of the models here is acceptable as much detail was provided before and the method is similar

$$k(x) = 0.033057x^3 - 1.1564x^2 + 12.513x + 1.8699$$

2.3 Modelling the First Guitar Body Lower Curve

Although the lower curve of the guitar body is not the same as the upper curve, it has similar characteristics as it has turning points in similar positions. Therefore, from my findings modelling the guitar body's upper curve, I used similar functions to model the lower curve. However, to avoid repetition, I will not be explaining the process.

Figure 2.3.1

Graph of the First Lower Curve

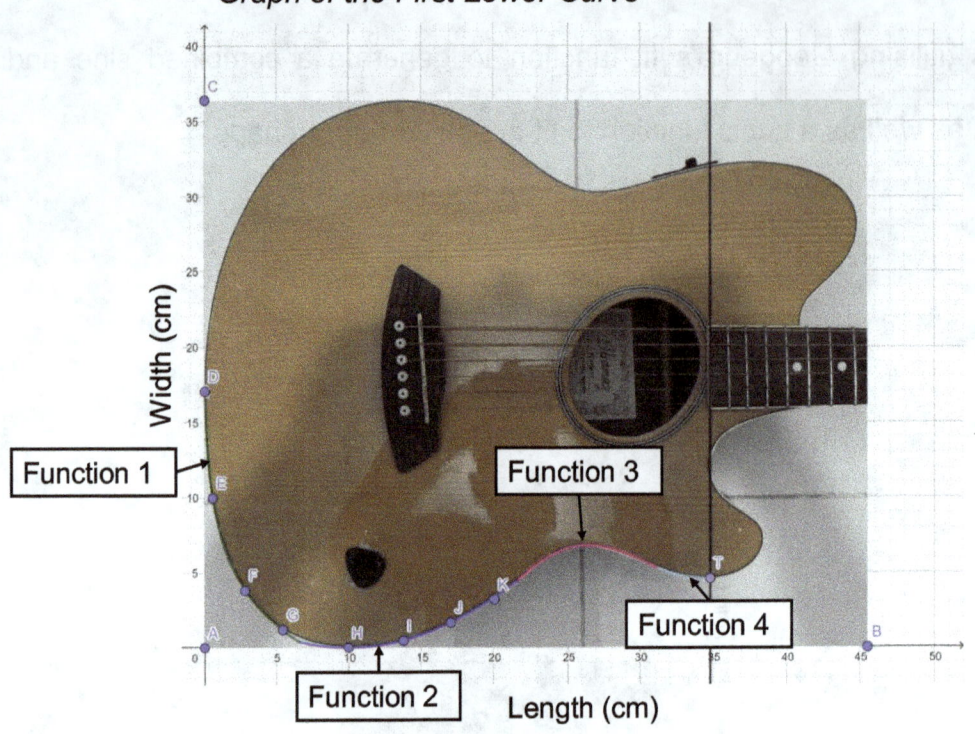

Function 1:   A+ B+ This page mk

$$m(x) = -4.1321 \ln(0.149x + 0.01779) + 3.8714$$

Function 2:

$$n(x) = 0.033(x - 10)^2$$

Function 3:

$$p(x) = 3.93 \sin(0.23x - 4.05) + 0.4x - 7.32$$

Function 4:

$$q(x) = 0.5\,(x - 34.805)^2 + 4.576$$

2.4 Modelling the Second Guitar Body Lower Curve

The second lower curve of the guitar horn is significantly more pronounced than the upper curve of the guitar horn, so I attempted to use GeoGebra's polynomial fit function to model the entire second lower curve, as seen below.

Figure 2.4.1

Graph of Cubic Function Modelling the Second Bottom Curve

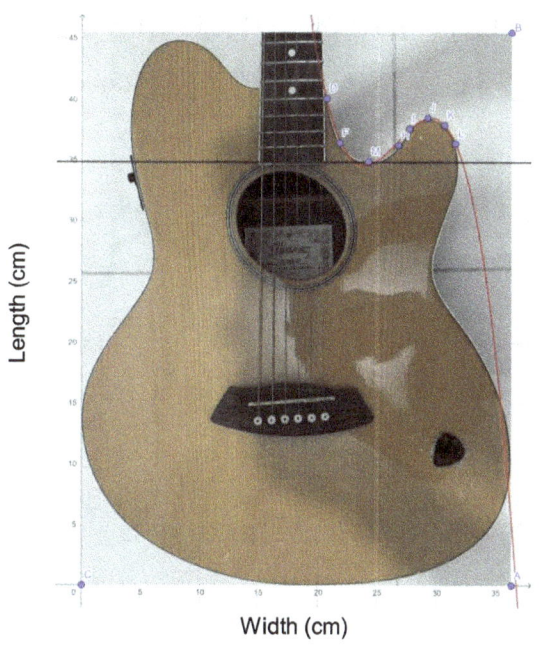

As the function approaches zero, it rises steeply to fit the guitar horn's shape. The fit function accommodates this rate of change but creates a turning point below the black line, which is not meant to intersect with the fit function because it is the boundary line that separates the guitar horn curve from the lower curve. Furthermore, the function misses parts of the guitar body. Hence, I decided to model the steep rise of the function separately as an exponential function and re-fit the cubic equation to fit the guitar body better.

> D+ A+ C+ Again, well written, really good engagement and reflection. Student is thinking about how move forward.

Figure 2.4.2

Graph of Cubic and Exponential Functions Modelling the Second Lower Curve

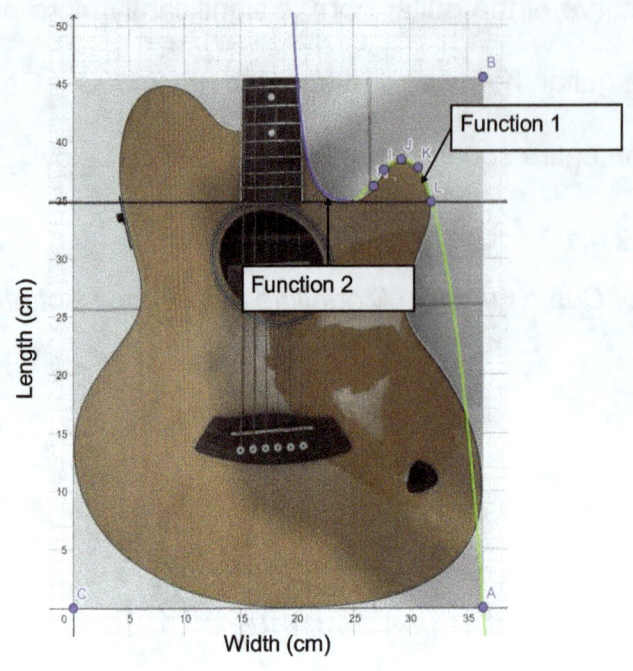

Function 1:

$$r(x) = 23.915\, e^{-1.0868(x-19.365)} + 34.827$$

Function 2:

$$s(x) = -0.050153x^3 + 4.0049x^2 - 105.50x + 952.83$$

It is important to note that the functions to model the second lower curve intersect, and this point of intersection is a sharp point, which means function 2 is not a smooth continuation of function 1. However, while the model is imperfect, it will only affect the area minimally as it is approximately overestimating less than 1 cm².

3.0 Calculating the Area of the Guitar Body's Soundboard

To find the area of the Guitar Body, the first step was to calculate the area of the upper curves.

3.1 Calculating the Area of the Upper Curve

To maintain conciseness, I will integrate the natural logarithmic upper curve function analytically and the rest using my GDC. I integrated the upper curve's first function, the ln function. I round my calculations to 6 d.p to maintain a high level of accuracy, but when I am adding the values of my calculations to determine the area of the guitar body, I will round to 5 s.f to enhance the clarity and readability.

Figure 3.1.1

Annotated Graph showing Area under the Upper Curve Natural Logarithmic Function

Clear diagram B+

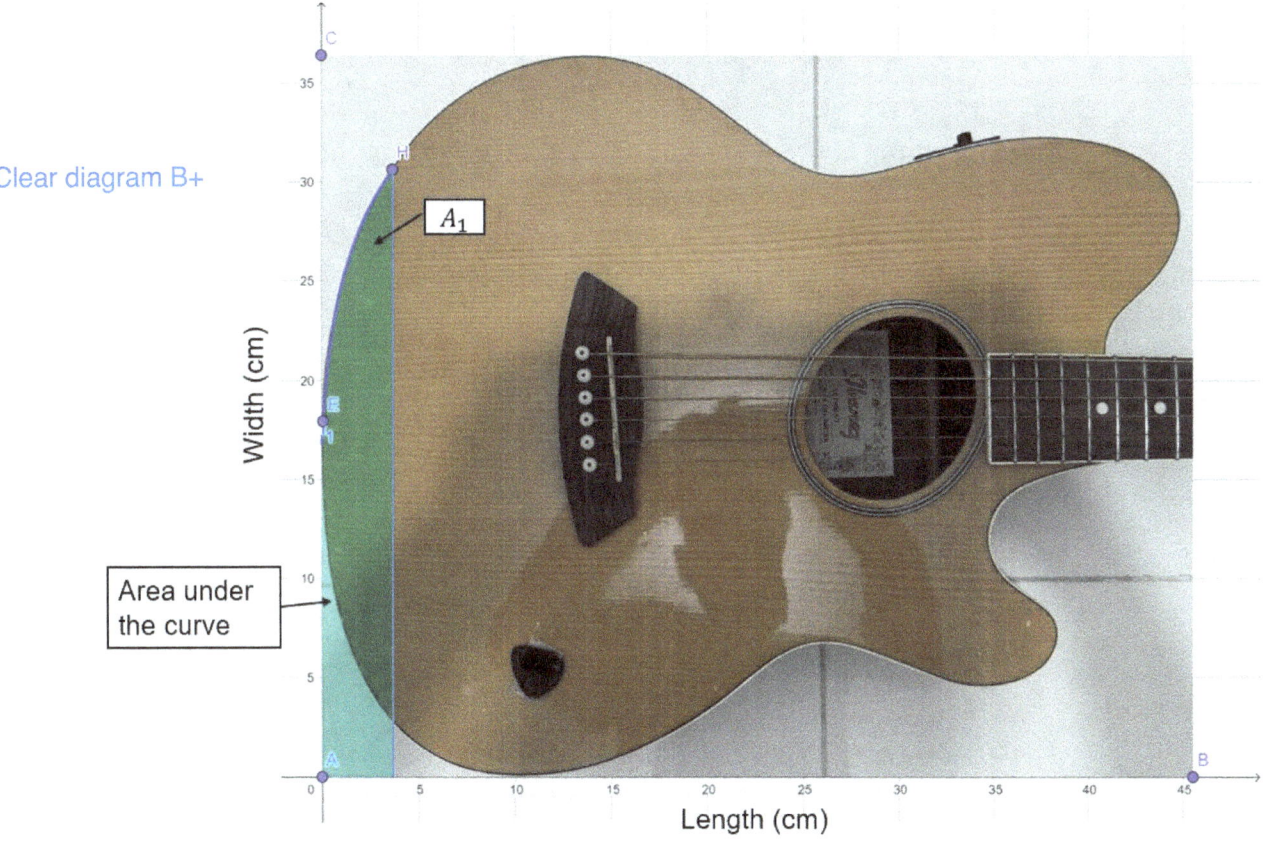

Integral:

$$A_1 = 5.2 \int_0^{3.6514} \ln(102x + 28.1)\, dx - \int_0^{3.6514} 0.57\, dx$$

$$= 5.2 \int_0^{3.6514} \ln(102x + 28.1)\, dx - 2.081298$$

To integrate the first integral, I will be using the integration by parts method:

$$= 5.2 \int_0^{3.6514} \ln(102x + 28.1) \, dx - 2.081298$$

$$u = \ln(102x + 28.1) \qquad\qquad \frac{dv}{dx} = 1$$

$$\frac{du}{dx} = \frac{102}{102x + 28.1} \qquad\qquad v = x$$

Hence:

$$= 5.2 \left(\left[x \ln(102x + 28.1) \right]_0^{3.6514} - \int_0^{3.5614} \frac{102x}{102x + 28.1} \, dx \right) - 2.081298$$

$$= 113.787363 - 5.2 \int_0^{3.5614} \frac{102x}{102x + 28.1} \, dx - 2.081298$$

The integral above can be solved using substitution. Therefore let:

$$u = 102x + 28.1$$

$$\frac{du}{dx} = 102$$

B+ E+ Good mathematical communication and a good level of sophistication and rigor demonstrated. Student understands integration by parts and explains it well

$$du = 102 \, dx$$

$$x = \frac{u - 28.1}{102}$$

Substitution of values into the integral:

$$= 111.706065 - 5.2 \int_0^{3.5614} \frac{\left(\frac{u - 28.1}{102}\right)}{u} \, du$$

$$= 111.706065 - \frac{5.2}{102} \int_0^{3.5614} \frac{u - 28.1}{u} \, du$$

$$= 111.706065 - \frac{5.2}{102} \left[u - 28.1 \ln|u| \right]_0^{3.6514}$$

$$= 111.706065 - \frac{5.2}{102} \left[102x + 28.1 - 28.1 \ln|102x + 28.1| \right]_0^{3.6514}$$

$$= 111.706065 - \frac{5.2}{102} \times 297.779665$$

$$= 96.5251411 \, cm^2$$

I will not show all the integration analytically because it would take away from the conciseness of the IA, so I used my GDC to integrate the rest of the upper curve functions.

Figure 3.1.2

Annotated Graph showing Area under the Upper Curve Functions

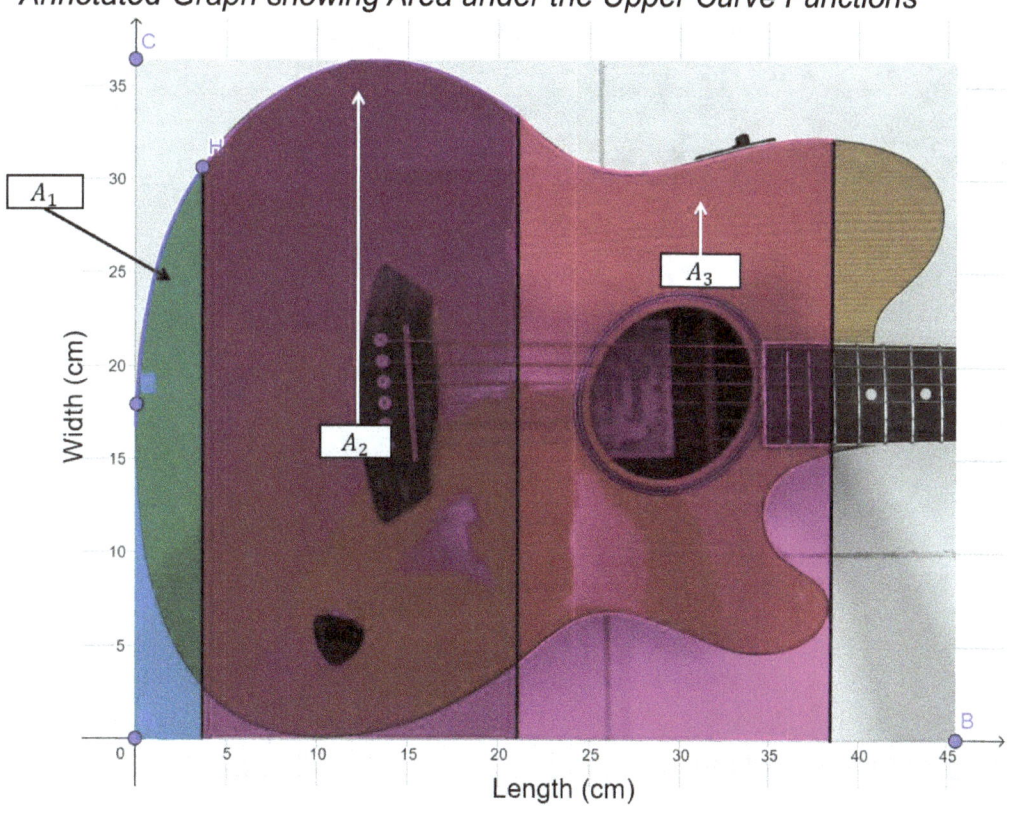

A+ B+ again, presentation is good and diagram is clear and helps with understanding

Cubic Equation 1:

$$A_2 = \int_{3.6514}^{21.056} (7.1564 \times 10^{-4})x^3 - 0.082640x^2 + 1.8245x + 12.047 \, dx$$

$$= 381.281638 \text{ cm}^2$$

Cubic Equation 2:

$$A_3 = \int_{21.056}^{38.565} -0.0035146x^3 + 0.34278x^2 - 10.873x + 143.29 \, dx = 549.733889 \text{ cm}^2$$

For the following functions, the values of the integrals need to be subtracted by the area below $y = 38.565$ as the previous functions have already accounted for it.

Figure 3.1.3

Annotated Graph showing Area under the Upper Guitar Horn Functions

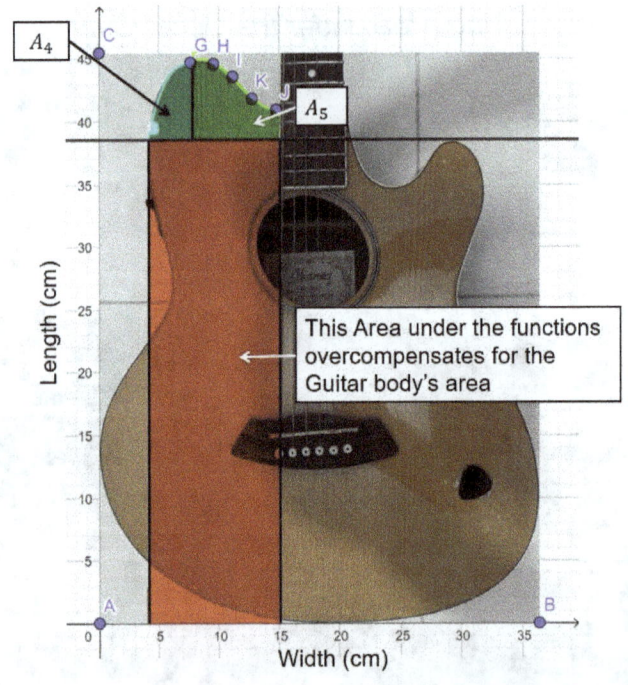

Upper guitar horn function 1:

$$\int_{4.0863}^{7.6527} -9842.75 \sin(0.041738x + 5.9609) + 395.97x - 3012.8 \, dx = 252.400608$$

$$A_4 = 252.400608 - (38.565 \times (7.6527 - 4.0863)) = 114.862392 \text{ cm}^2$$

Upper guitar horn function 2:

$$\int_{7.6527}^{15.070} 0.033057x^3 - 1.1564x^2 + 12.513x + 1.8699 \, dx = 319.751714$$

$$A_5 = 319.751714 - (38.565 \times (15.070 - 7.6527)) = 33.703540 \text{ cm}^2$$

Total upper curve area:

$$96.5251411 + 381.281638 + 549.733889 + 114.862392 + 33.703540$$

$$= 1176.1066001 \text{ cm}^2$$

3.2 Calculating the area of the Lower Curve

Again, I will analytically integrate one function of the lower curve and calculate the rest using my GDC.

Figure 3.1.4

Annotated Graph showing Area under the Lower Curve Guitar Functions

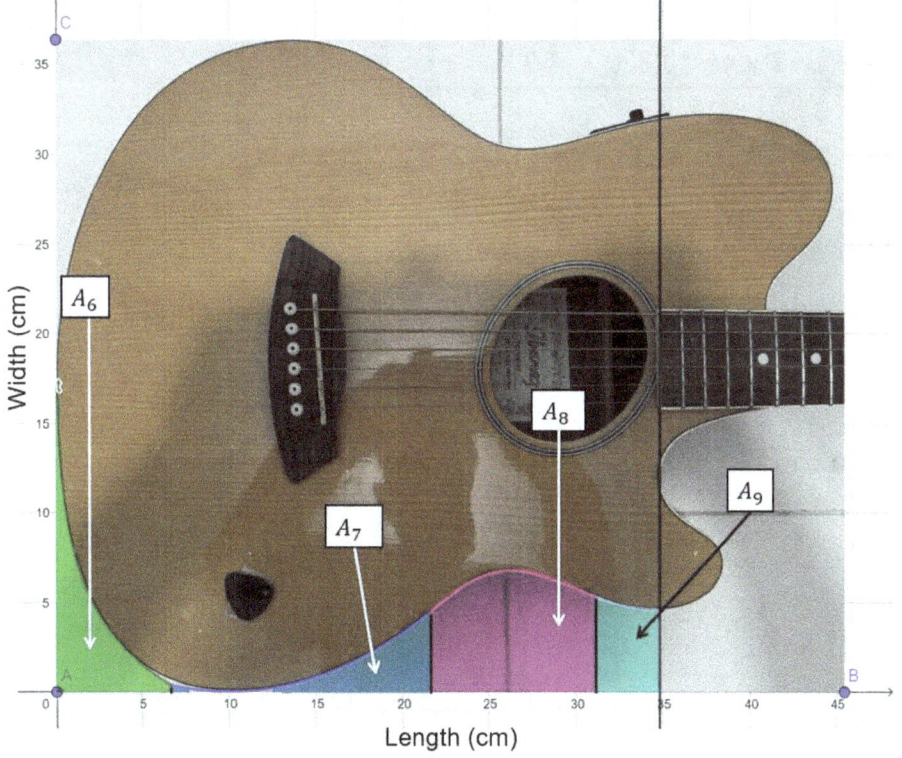

Lower curve ln function:

$$A_6 = \int_0^{6.6017} -4.1321\ln(0.149x + 0.01779) + 3.8714 \; dx = 50.808861 \text{ cm}^2$$

Lower curve quadratic function:

$$A_7 = \int_{6.6017}^{21.592} 0.033(x-10)^2 = 17.566052 \text{ cm}^2$$

The lower curve function I will be integrating analytically is the lower curve's sine function:

$$A_8 = \int_{21.592}^{31.168} 3.93\sin(0.23x - 4.05) + 0.4x - 7.32 \; dx$$

$$= \left[-\frac{3.93 \times \cos(0.23x - 4.05)}{0.23} + \frac{0.4x^2}{2} - 7.32x\right]_{21.592}^{31.168}$$

$$= \left[-\frac{393 \times \cos(0.23x - 4.05)}{23} + \frac{x^2}{5} - 7.32x\right]_{21.592}^{31.168}$$

$$= 58.435837 \text{ cm}^2$$

Lower curve second quadratic function:

$$A_9 = \int_{31.168}^{34.805} 0.5\,(x - 34.805)^2 + 4.576 = 24.661145 \text{ cm}^2$$

Total lower curve area:

$$50.808861 + 17.566052 + 58.435837 + 24.661145 = 151.471895 \text{ cm}^2$$

Like the functions of the upper guitar horn, integrals of the bottom guitar horn functions need to be subtracted by the area below $y = 34.805$.

Figure 3.1.5

Annotated Graph showing Area under the Lower Guitar Horn Functions

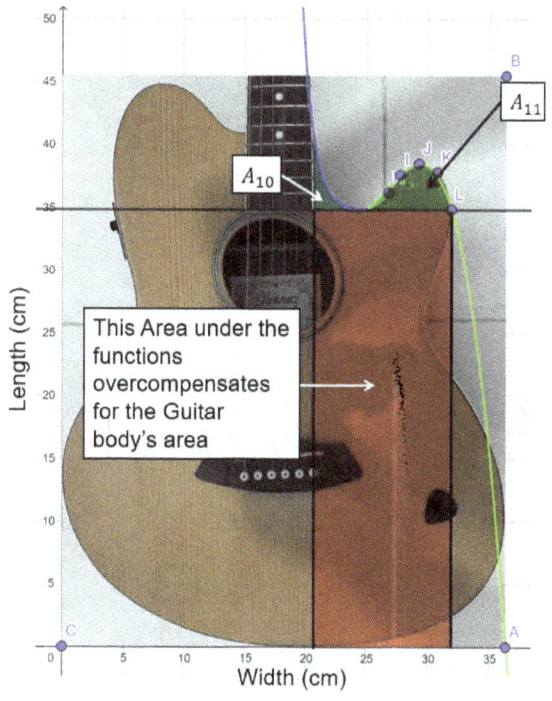

Lower guitar horn function 1:

$$\int_{20.546}^{25.022} 23.915 \, e^{-1.0868(x-19.365)} + 34.827 = 161.935372$$

$$A_{10} = 161.935372 - (34.805 \times (25.022 - 20.546)) = 6.148192 \text{ cm}^2$$

Lower guitar horn function 2:

$$\int_{25.022}^{31.864} -0.050153x^3 + 4.0049x^2 - 105.50x + 952.83 \, dx = 252.862250$$

$$A_{11} = 252.862250 - (34.805 \times (31.864 - 25.022)) = 14.726440 \text{ cm}^2$$

Total area of lower guitar horn:

$$6.148192 + 14.726440 = 20.874632 \text{ cm}^2$$

3.3 Calculating the Area of Sound hole and Guitar Neck

I recognized that the shape of the sound hole looked like an ellipse. So, I measured the appropriate values to calculate the area using the ellipse area formula:

Figure 3.3.1

Image showing Area of an Ellipse Formula

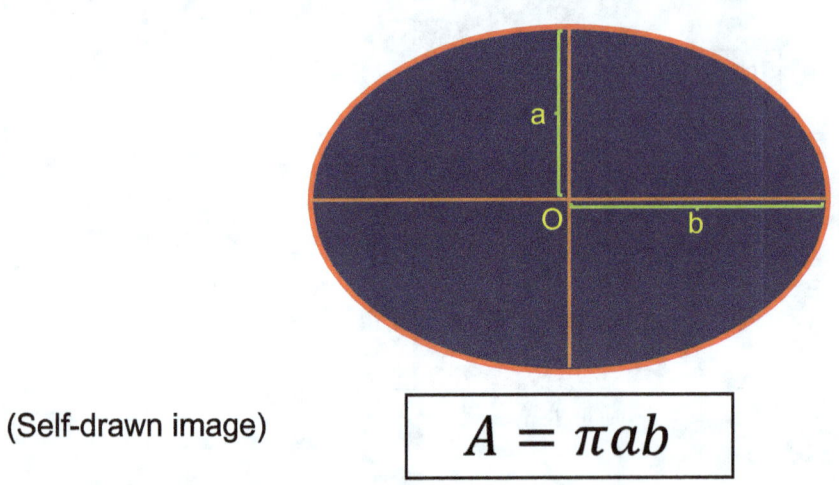

(Self-drawn image)

The general formula for the area of a circle is πr^2 however, an ellipse has two different radii, so the formula multiplies pi by the two radii. Still, it should be noted that this formula can be proved through integration.

Using this formula, I calculated the area of the sound hole using my measured values:

$$A = \pi \times 4.25 \times 5 = 66.758844 \text{ cm}^2$$

Moving on to calculate the area of the guitar neck I also measured its length and width to determine its area:

$$A = 6.5 \times 5.6 = 36.4 \text{ cm}^2$$

3.4 Calculation for the Guitar Body's Area

The guitar body's area can be calculated by adding the area calculated from the upper curves and the lower guitar horn and then subtracting the area calculated from the lower curves, the sound hole, and the guitar neck.

$$1176.1066001 + 20.874632 - 151.471895 - 66.758844 - 36.4 = 942.350493$$

$$\approx 942.35 \text{ cm}^2$$

The calculated area appears correct as it aligns with the guitar's dimensions of a 36.4 cm maximum width and 45.5 cm maximum length, and the product of these values is 1656.2 cm². However, considering the irregular shape of the guitar body and its significant cutaways, a calculated guitar soundboard area of 942.35 cm² seems valid.

4.0 Calculating the Volume of Guitar Wax Needed

The volume needed for one layer of guitar wax can be calculated by multiplying the area of the guitar body by the assumed thickness of guitar wax (0.2mm or 0.02 cm).

$$942.35 \times 0.02 = 18.847 \text{ cm}^2$$

Given that I wax my guitar once a month, over a year I would wax the guitar body's soundboard 12 times.

Hence, the total volume of guitar wax I use is:

$$18.847 \times 12 = 226.164 \approx 220 \text{ cm}^2 \text{ or } 220 \text{ ml}$$

I round my answer to 2 s.f as it gives a more reasonable estimate. Therefore, I would use approximately 220 ml of guitar wax in one year to wax my guitar's soundboard.

A+ D+ good conclusion. More reflection. Nice ending to a very well written Exploration.

5.0 Conclusion

The investigation aimed to calculate the volume of guitar wax I use on my guitar's soundboard during a year. I was successful in doing so by modelling my guitar's body using multiple functions and integrating them to get a value for the area. Then, by subtracting the area of the guitar's sound hole and guitar neck from the integrated value for the area, I got the area of the guitar's soundboard. Then, multiplying it by the assumed thickness of a guitar wax layer, I found the volume of guitar wax needed for one use and, therefore, for one year. Hence, I would need about 220 ml of guitar wax yearly to wax my guitar's soundboard.

However, this is based on the assumption of guitar wax thickness, which can be variable, creating uncertainty about whether my calculations give a valid estimate of how much guitar wax will be required.

Interestingly, the functions used fit the guitar body's shape well, supporting the validity of the conclusion that around 220 ml of guitar wax will be used to wax my guitar soundboard for a year. However, uncertainty is introduced as areas of the shape are either missed or cut into the guitar body by the functions, although very minimal. Considering that a build-up of slight errors can lead to an overestimate of the volume of guitar wax needed, investigating further functions like parametric equations could improve the investigation as they are flexible in adjusting the function's characteristics. To conclude, further functions can be tested to model the shape of the guitar body more accurately.

Bibliography

Cuemath. (n.d.). *Area of ellipse*. Cuemath. Retrieved October 30, 2024, from https://www.cuemath.com/geometry/area-of-an-ellipse/

Guitar wax - page 4 - the acoustic guitar forum. (2014). Acousticguitarforum.com. https://www.acousticguitarforum.com/forums/showthread.php?t=324296&page=4

Lumen. (n.d.). *Graphing transformations of logarithmic functions*. Courses.lumenlearning.com. Retrieved October 19, 2024, from https://courses.lumenlearning.com/ivytech-collegealgebra/chapter/graphing-transformations-of-logarithmic-functions/

Mathematics Libre Texts. (n.d.). *Graphs of sinusoidal functions*. Math Libre Texts. Retrieved September 21, 2024, from https://math.libretexts.org/Bookshelves/Precalculus/Book%3A_Trigonometry_(Sundstrom_and_Schlicker)/02%3A_Graphs_of_the_Trigonometric_Functions/2.02%3A_Graphs_of_Sinusoidal_Functions

OntarioTech University. (n.d.). *Transformation of exponential and logarithmic functions*. Nool.ontariotechu.ca. Retrieved October 19, 2024, from https://nool.ontariotechu.ca/mathematics/exponential-logarithmic-functions/basics/transformation-of-exponential-and-logarithmic-functions.php

Uitti, J. (2023, July 17). *The 20 best Eddie Van Halen quotes*. American Songwriter. https://americansongwriter.com/the-20-best-eddie-van-halen-quotes/

Yamaha. (n.d.). *How the acoustic guitar is made: Coating and polishing gives the instrument a beautiful finish - musical instrument guide.* Www.yamaha.com. Retrieved October 17, 2024, from https://www.yamaha.com/en/musical_instrument_guide/acoustic_guitar/manufacturing/manufacturing004.html

IA Sample 9 Examiner Comments: Triathlon Progress

This IA is a good example of a student-led investigation into a meaningful and personal context. The student uses statistics to investigate her triathlon performance.

Appropriate for: AASL, AISL

Possible for: AAHL, AIHL (You could add some of the HLAI tests)

Criterion	SL	HL
A	3	3
B	3	3
C	3	3
D	2	2
E	5	4
Total	16	15

A: The IA is logically structured with two clearly stated aims. Each aim is followed through in its own section. Graphs and tables are relevant and support understanding. It is an easy read and easy to follow. It flows well although two aims take from the conciseness, hence the dropped mark.

B: Mathematical processes such as regression and t-test calculations are communicated with generally clear working and notation. Variables are defined. Axes on some graphs are not labelled, and calculator screenshots are not appropriate, hence the dropped mark.

C: This IA is highly personal, as it explores the student's own running data and connects directly to their triathlon training. The motivation is authentic, and the investigation feels owned by the student. The mathematics guides the exploration. An example of this is where the student reflected on their linear model and considered creating an exponential model.

D: The student reflects on the meaning of their regression fits and interprets the t-test. They discuss how the models fit the data and what the comparison tells them about fatigue. Not enough critical reflection to warrant D3.

E: The student applies two different mathematical tools: regression (linear and exponential) and t-tests executed correctly. The regression work includes interpreting R^2 values and using equations for prediction, while the t-test includes formulating hypotheses and drawing conclusions based on the test statistic. The maths is used to meaningfully explore both aims. Good but not thorough understanding is demonstrated with regards to the exponential model.

Contents

No need for a Table of Contents

Introduction 2
Aims 2
Rationale 2
Product Moment Correlation Coefficient 4
Linear Regression Equation 5
    X on Y Regression Line 5
Coefficient of Determination 6
Exponential Model 7
Aim 2 9
    Raw Data 9
Descriptive statistics 10
    Outliers 10
    Mean 12
    Standard Deviation & IQR 12
T - Test 14
Conclusion 15
Bibliography 16
Appendix 17

A - Introduction, aim, rationale and conclusion included.

A - this investigation is organised and logically structured.

A - IA is divided into clear, sensible, signposted subsections.

Introduction

Throughout my life, I have always loved sport. Originally, I was a swimmer but switched to triathlon three years ago. Triathlon is a sport that consists of swimming, cycling and running; through its increasing global recognition, more and more people have decided to take up the sport, subsequently resulting in a plethora of local races. Fortunately, this enabled me to compete more regularly and therefore, understand what disciplines I need to improve. Over the past 3 years, I have taken part in numerous local races as well as some international races which included representing Great Britain in the 2022 Age Group World Championships. Thus, I am interested to see how my performance has changed over time, particularly the rate at which I've improved. By training consistently over the past few years and my body adapting to become stronger and fitter with age, I believe my times have gotten quicker but, am very curious to see the specific details by conducting this investigation. As I have been racing since 2020, I have collected official times from all the races I participated in.

C - student will be collecting their own data and addressing a personal interest.

Aims

First Aim: To investigate the relationship in my performance over time.

Second Aim: Compare my 5km running times in a triathlon to my 5km running times in a running race to see how fatigue affects my performance.

A - explicit aims.   *A - aims met through investigation.*

Rationale

I have decided to investigate the relationship between age and performance as I am interested to measure the rate of change in my performance over time. As a whole, athletes are obsessed with their times and overall performance. We dedicate our lives to the sport and thus, our personal performances are of huge importance to us. Therefore, it is vital for me to see my rate of improvement in triathlon. This will give me an insight into whether my training schedule and load has enabled me to progress as a triathlete.

By conducting this investigation, I will not only understand the rate of change in performance, but also be able to see whether adapting my training loads is necessary. Triathlon is a discipline that I am passionate about and by further developing my knowledge on my progress and subsequent data trend, it will allow me to become a better informed athlete. I want to conclude, what the rate of change of my performance over time is, and understand how fatigue affects my performance by comparing my 5km running within and away from a triathlon.

C - personal investment in the outcome of the aim.

> **Reflection:**
> I believe I will find a positive correlation between age and performance, thus meaning as training duration increases, my performance level has increased. Through changes in my body composition, size and height, coupled with my dedication to the sport, I should be producing quicker results.

C - student is considering the impacts of the investigation and their findings.

C - looking forward.

By finding all my race times online since 2020, I have gathered all data I can access from these races, from organisations such as Sport Sports and Race M.E. I believe this data is reliable as it has been taken from external providers who produce precise and accurate data; all performers are timed with advanced technology, the ProChip system, and individual athletes are timed with a timing chip. This highlighting the reliable nature of this data for my investigation. However, there may be some limitations to this information as, due to me having only done the sport for three years, I am limited on the amount of data I can collect.

Raw Data   *D - considering reliability and limitations*

Table 1.1 shows the days since starting triathlon and the respected time it took to complete the triathlon.

Days since starting triathlons (x)	Total time taken for the triathlon (seconds) (y)
0	4676
7	4544
32	4488
208	4721
287	4370
372	4391
455	4408
644	4146
651	4065
738	4066
750	4064
759	3950
772	4018
828	4089
911	4159
1017	3973
1042	3926
1049	3784
1101	3894
1121	3871
1157	3837
1192	3756

A - clear and concise table.

B - units appropriate and defined.

Reflection:
To increase reliability and enable a trend to be found, I ensured the total time taken was rounded to the nearest second. However, varying environmental conditions, such as terrain and wind create issues with accuracy of the time produced. Without controlling the environment, it limits the data's reliability as the times are not produced in the same conditions and therefore, the comparison between the performance's may be less valid. Moreover, the external providers do not track the information to the millisecond which further hinders my investigation. However, with all results being recorded with technology, we can suggest the times do have some validity because they avoid human error.   *D - looking backwards and considering limitations*

E - has demonstrated a thorough understanding of the logical sequence of mathematical approaches.

B - graph clearly labelled and consistent with table of data, relevant representation to support aim 1.

A scatter graph can be used to study my performance trend and model the respected data. This enables me to visually understand the relationship between the two variables, total time taken and days since starting, before moving onto specific calculations.

Graph 1.2 showing the relationship between days since starting triathlons and total time taken

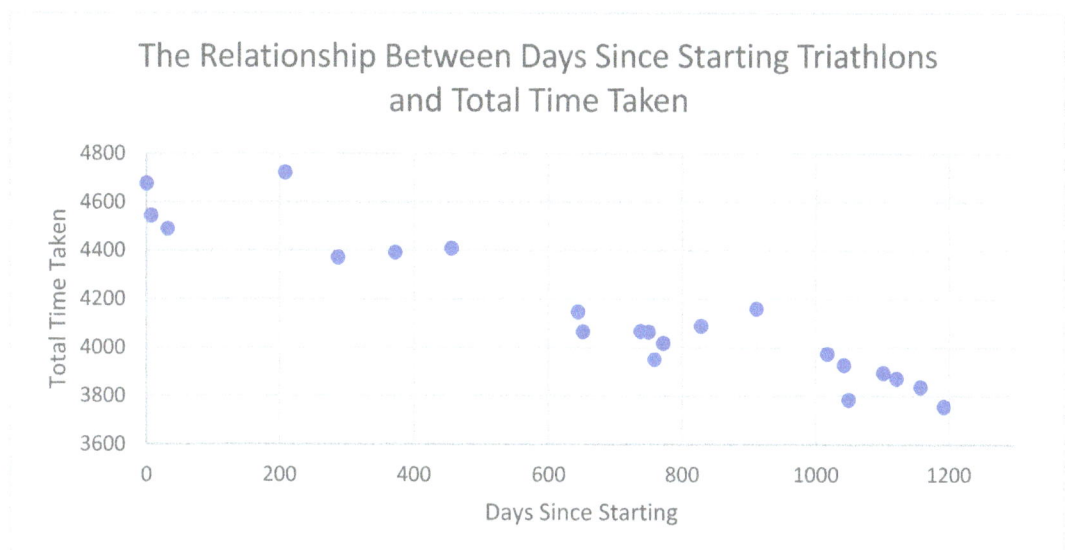

Product Moment Correlation Coefficient

By calculating the Product Moment Correlation Coefficient, the r value, I will be able to determine the strength and type of correlation between days since starting triathlon and total time taken. From looking at the graph, I believe this will be a negative correlation however, I need to find the r value to determine the relationship's strength. I am able to find the r value as the data forms a linear pattern. The table below indicates the r values needed to conclude the correlation's strength.

D - excellent, meaningful reflection on how to analyse correlation coefficient.

Product Moment Correlation Coefficient	Level of Correlation
$0.0 \leq \|r\| < 0.20$	Negligible/None
$0.20 \leq \|r\| < 0.40$	Weak
$0.40 \leq \|r\| < 0.65$	Moderate
$0.65 \leq \|r\| < 1$	Strong
$\|r\| = 1$	Perfect

Through a calculation on Edexcel, I found the r value as $r = -0.95$ to $3\ sf$. Thus, there is a very strong negative linear correlation between days since starting triathlons and total time taken, suggesting as days since starting increases, the time taken to complete a race decreases.

D - critical reflection of r-value and its implications.

B - excellent use of technology.

Reflection:

Initially, I was expecting a negative r value with a relatively strong correlation; a negative r value indicates an improvement in my time and I believe my performance has positively progressed in a linear relationship over time. However, I did not think the relationship between the two variables would be so strong. Performance is a fluctuating, unpredictable variable, hence why I was expecting a smaller correlation. However, in this time, I've gained more experience and physical fitness highlighting the r values validity.

C - student voice can be heard.

y on x Linear Regression Equation

After finding my r value, I will now plot a line of best find to determine the y on x regression equation. With the data having strong negative correlation, I will visually represent the relationship using a linear regression equation.

Graph 1.3 containing the line of best fit and the y on x linear regression equation.

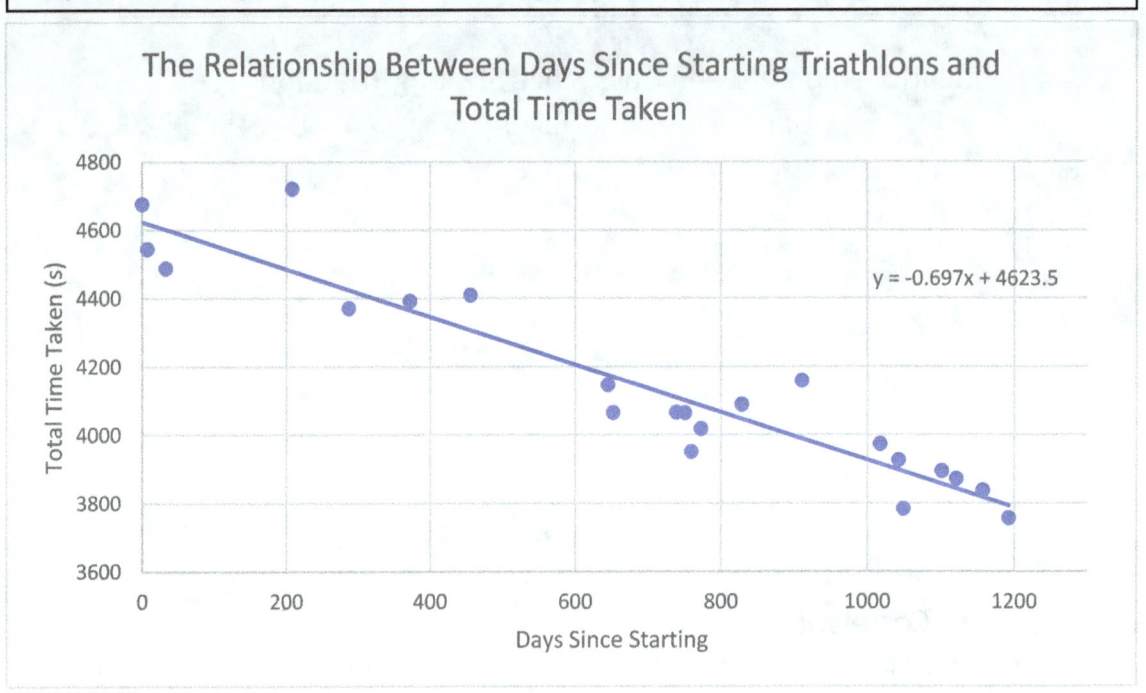

A - communication is concise.

Y on X Regression line: $y = mx + b$

B - variable concisely explained.

I used excel to find the linear regression line as $y = -0.697x + 4623.5$ where y represents the total time taken and x represents days since starting triathlon. The x coefficient represents the gradient; with its value being $m = -0.697$, this demonstrates that an increase in 1 day since starting will result in a time being 0.697 seconds quicker. After completing 100 days of training, the gradient suggests my subsequent time will be 69.7 seconds faster. The y intercept is representative of my time following 0 days of training, implying my baseline line time to complete a triathlon would be 4623.5 seconds.

E - relevant mathematics and commensurate with the SLAI course. Correctly calculated and reflected upon results.

> **Reflection:**
> I believe this regression line is reliable. Through training consistently, my time taken should decrease as my cardiovascular endurance will increase. Moreover, these values are reliable as 100 days since training lies within my data range.

x on y Regression Line

By changing my linear regression line to a x on y regression line, it enables me to find how many days I have to train to perform a specific time.

C - evidence of student interest, mathematics beyond the course.

B - excellent use of technology and understanding demonstrated.

Calculator screenshots are not recommended

	=LinRegM
Title	Linear R...
RegEqn	m*x+b
m	-1.28199
b	6000.25

X on Y Regression Line: $x = my + b$

Through calculations on my GDC, I found the x on y regression line as $x = -1.282y + 6000$ where y represents the total time taken and the x represents the number of days of training.

Figure 1.4 shows the winning time at WTS Leeds in 2022 as 59:03 seconds

Pos	First Name	Last Name	YOB	Country	Start Num	Time	Swim	T1	Bike	T2	Run
1	Cassandre	Beaugrand	1997	FRA	18	00:59:03	00:09:09	00:01:11	00:32:13	00:00:22	00:16:10

In order to understand my performance level greater, I have decided to see how many days I need to train to hit a winning World Triathlon Series time. Figure 1.4 shows the winning time at the WTS Leeds 2022; being English, winning a WTS race in my home country is an aspiration of me and therefore, I wanted to see whether this is possible with my current progress trend. To win this race in 2022, as shown above, I would need to have a total time of $y = 3543$ seconds.

D - very important note, considers further investigation.

X on Y Regression Line calculations:

$$x = -1.282(3543) + 6000$$

$$x = 1458$$

By using my linear regression x on y line, it suggests I would need to train for 1458 days. With this being achievable, it motivates me to continue training, implying I could potentially hit this time after another 266 days of training. However, despite this being very achievable, its reliability is limited.

D/E - thorough knowledge and understanding demonstrated.

Reflection :

Through the x on y regression line, I can see that I have a very high potential in becoming an elite triathlete however, I cannot completely conclude this is a valid statement. By trying to hit a specific time that is not within the data range, I am unable to predict if this trend will continue. Moreover, performance cannot continue to increase indefinitely and must plateau at one point otherwise this would result in continual world records. Thus, I must conclude with caution and understand the days since training may be invalid.

D - meaningful reflection and limitations of the math considered.

Coefficient of Determination

Following on, I will now find the coefficient of determination. This will measure how well my statistical model above predicts its respected outcome. Essentially, it will enable me to understand what percentage of days trained impacts the subsequent times achieved.

C - evidence of student interest, mathematics beyond the course.

B - appropriate mathematical communication and notation.

Coefficient of Determination, r^2	Level of Influence
$0.82 \leq r^2 \leq 1.00$	Very strong
$0.48 \leq r^2 \leq 0.82$	Strong
$0.17 \leq r^2 \leq 0.48$	Moderate
$0.05 \leq r^2 \leq 0.17$	Low
$0.00 \leq r^2 \leq 0.05$	Very low

Excel found the coefficient of determination value as $r^2 = 0.8936$, implying a very strong level of influence between days trained and total time taken in a triathlon. The r^2 value suggests approximately 90% of all times can be explained by my continuous efforts in training, highlighting training vast benefits for my performance. By training consistently, I am improving my aerobic and anaerobic capacities as well as speed components; these evidently play a huge part in decreasing my total time taken, shown by the high r^2 value. However, although the r^2 value calculated does demonstrate a very strong level of influence, with the value not being 1, it suggests there are still areas that are not completely interdependent. In a triathlon, extrinsic variables such as the wind conditions, weather and terrain will affect the total time taken. Moreover, intrinsic factors such as motivation, arousal, sleep and nutrient will play a part in my respected time. Thus, it is perhaps inevitable that days trained and total time taken do not completely influence one another, despite having a significant influence.

E - has shown an understanding of mathematical concepts and applied them correctly.

Reflection:
Although I recognise there is a linear relationship and a high influence between days trained and time taken, a linear model has its limitations. A linear graph does not plateau and will eventually reach 0 seconds, hindering its reliability for understanding a performance due to the impossibility of performing a triathlon in 0 seconds. Thus, I will now use a different form of modelling to make my data trend line more reliable in a real life context.

D - critical reflection on accuracy of own findings.

C - questioning their own findings and understanding mathematical limitations.

Exponential Model
C - student has allowed the mathematics to drive the investigation forward.

I believe an exponential model will prove a better model for my long term performance. Unlike a linear model, an exponential graph will plateau as it does not cross the x-axis thus seeming more reliable. Moreover, it has a gradual curve reflecting my rate of change in performance more accurately.

Exponential regression equation: $y = ab^x$

B - Clear representation of formulae.

By using my GDC I have found my exponential regression as $y = 4634 \times 0.9998^x$ to 4 sf.

y	C	D	E
			=ExpReg(
1	4676	Title	Exponen...
2	4544	RegEqn	a*b^x
3	4488	a	4633.68
4	4370	b	0.999834
5	4721	r²	0.896207

Key:
$a = y$ intercept – initial total time taken
$b = $ constant multiplier (the rate of change)
$x = $ days since starting
$y = $ total time taken

B - defines key variables.

Graph 1.5 showing the exponential regression line.

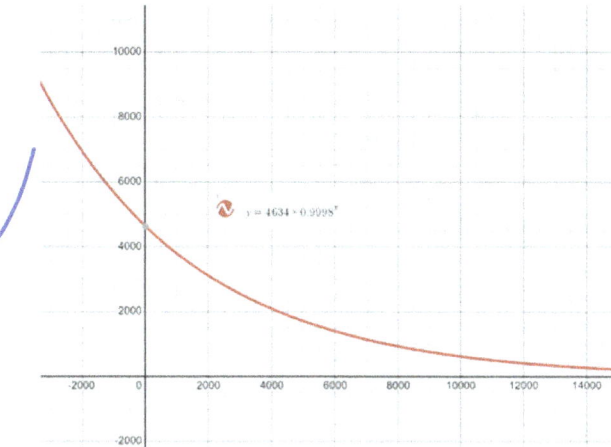

Modelling my trend exponentially demonstrates my rate of change of performance more reliably. The a value is synonymous with the y intercept, representing my initial performance time as $a = 4634$. The b value suggests my rate of change overtime, implying as day since starting increases by 1, the total time taken will be a will be a product of the base, and the multiplier value of $b = 0.9998$, resulting in a decrease in time.

E - good knowledge and understanding demonstrated.

Reflection:

I believe an exponential model is a more reliable representation my rate of change in performance. With it being shaped as a curve rather than a linear path, it mirrors my performance's progress more accurately. Despite progressing through time, my rate of progress can be minor. As you get older and slower, it becomes more challenging to continually hit a personal best. An exponential model reflects this, portraying significant initial progress but then levelling off, suggesting a time improvement becomes more difficult, alluding to similar characteristics of an athletes progress. Thus, this model is more beneficial for the representation of sports progression.   A - focus is on the aim and finding the best model/relationship.

Figure 1.6 showing the Sprint Triathlon World Record Time

Pos	First Name	Last Name	YOB	Country	Start Num	Time	Swim	T1	Bike	T2	Run
1	Katie	Zaferes	1989	USA	2	00:55:31	00:09:08	00:00:55	00:28:55	00:00:25	00:16:09

Following this, I have found the current Sprint triathlon world record time. This was performed by American triathlete, Katie Zaferes in 2019, World Triathlon Series Abu Dhabi. As shown above, she completed the triathlon in 55: 31 minutes. By gaining this data, I have converted the time into seconds, and placed another equation, $y = 3331$. This enables me to find out how many days I would need to train to hit the world record time based on my current performance trend.   C - applying model to a real-world relevant scenario.

B - excellent use of technology.

E - mathematics explored is correct.

Graph 1.7 showing the exponential regression line and $y = 3331$

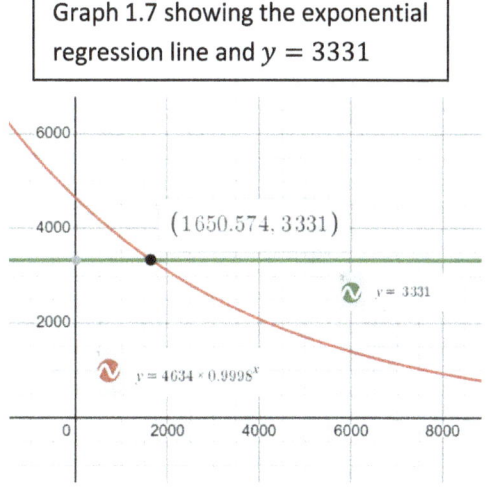

After adding the world record equation of $y = 3331$, the current trend line suggests I would need to train 1651 days to reach the world record time. Thus, I would need to train for approximately another 458 days to perform at world record pace. However, this form of modelling is not wholly reliable. Performance can sometimes plateau and improvements can be limited. Additionally, my prediction of training for another 458 days to reach world record pace is a prediction in the future; values to form this predications are outside of the data range suggesting extrapolation would have to take place which limits the reliability of my claim. Yet, despite these limitations the trend line, through its shape and characteristics, depicts my rate of change more accurately than a linear regression.   A - communication is concise.

D - excellent reflection for a student studying SLAI course.

Reflection:

Whilst I recognise that an exponential model shows my rate of change in performance more reliably than a linear curve, it also has its limitations. Exponential models are monotonic suggesting my trend line will continually decrease; performance fluctuates and my trend may not always be followed. Therefore, perhaps a quadratic, due to its minimum and maximum values, would help distinguish the trend greater, particularly in the long term as speed will hit an optimum value, then due to age, may start to decrease. However, as I have not reached my peak or slowed down yet, modelling with a quadratic would be invalid. Ultimately, to accurately demonstrate performance a combination of all model would be necessary. Sports performance is difficult to model due large number of variables that influences it, highlighting each model's inevitable limitations.

E - relevant mathematics and commensurate with the SLAI course. Correctly calculated and reflected upon results throughout, correctly linked to aim 1.

Aim 2

Following on from my first aim, I am going to compare my 5km running times in triathlons to my 5km running time in running races to see how fatigue effects my performance. I will be using descriptive statistics; calculating averages and measures of dispersion to demonstrate whether there is any differences between my two performances.

Raw Data

Table 2.0 showing my running performances in the respected races since 2020.

Run Time (seconds)(x)	Triathlon Run Time (seconds)(y)
1105	1180
1140	1206
1146	1212
1110	1170
1139	1089
1132	1190
1120	1214
1137	1221
1145	1291
1180	1252
1220	1200
1201	1265
1190	1268
1180	1278
1165	1228
1260	1354
1245	1302
1301	1357
1300	1404
1240	1250
1267	1354
1254	1320

A - clear and concise table.

Descriptive statistics

Outliers

By initially calculating outliers, I will eliminate the issue of data variability and increase the accuracy and statical power of the data. Outliers can cause significant impacts on the mean and standard deviation. Therefore, it is vital to mitigate there impact to increase the reliability of my investigation. Although, they do not affect the medium or mode, with my investigation being a comparison, the mean is seen as the most significant factor. Therefore, it is highly important to eliminate any outliers within the data.

Variables	Run Time	Triathlon Run Time
Q_1	1139	1206
Q_3	1245	1302
IQR	106	96

Quartiles are types of percentiles and divide the data into 4 percentiles with equal probability. The lower quartile, Q_1, represents the value that 25% of the data sits below while the upper quartile, Q_3, is known as the value that 75% of the data falls below. By finding these values, I can distinguish the Interquartile range. The IQR measures the statistical dispersion and is representative of the middle 50% of the data.

B - defines key variables.

$$IQR = Q_3 - Q_1$$

$$IQR\ Run\ Time = 1245 - 1139$$

$$IQR\ Run\ Time = 106$$

$$IQR\ Triathlon\ Run\ Time = 1302 - 1206$$

$$IQR\ Triathlon\ Run\ Time = 96$$

B - appropriate mathematical symbols used.

> **Reflection:**
>
> As shown above, the 5km run time values are quicker than the triathlon 5km run times indicated by a lower value for Q_1 and Q_3. However, the IQR value for the run time is greater. This suggests there is a greater spread of data, suggesting more variability. Initially, I believed the triathlon run time would have a larger IQR due to fatigue and varying motivation during the final stages of a triathlon. With the IQR being 50% of the data values, it does not take extreme values into account. Therefore it is valid to conclude that when running a 5km in a triathlon, my times are more consistent, altering my initial beliefs.

D - the student has reviewed their answers against their own predictions, in context of the aim.

Outlier formulas

Lower Limit $= Q_1 - (1.5 \times IQR)$

Upper Limit $= Q_3 - (1.5 \times IQR)$

B - Clear representation of formulae.

B - appropriate mathematical symbols used.

Run time:

Run time lower limit $= 1139 - (1.5 \times 106)$

Run time lower limit $= 980$

Run time upper limit $= 1245 + (1.5 \times 106)$

Run time upper limit $= 1404$

E - good knowledge and understanding demonstrated.

E - relevant mathematics commensurate with the SLAI course is logically explored.

E - mathematics explored is correct.

Triathlon run time:

Triathlon run time lower limit $= 1206 - (1.5 \times 96)$

Triathlon run time lower limit $= 1062$

Triathlon run time upper limit $= 1302 + (1.5 \times 96)$

Triathlon run time upper limit $= 1446$

Following these calculations, any run time outside the range of $980 \leq x \leq 1404$ and any triathlon run time outside the range of $1062 \leq y \leq 1446$ are considered outliers.

Reflection:

Despite doing these calculations, there are no extreme values within my data collection. This indicates that all my data points are consistent with the overall pattern of distribution. However, this calculation was still necessary to do; outliers effect the mean and may skew its value, proving detrimental for my investigation. Removing such values, make the comparison between run time and triathlon run times more accurate and reliable. Therefore, I can now proceed to calculate the measures of desperation accurately and understand the relationship between run times and triathlon run times in more depth.

D - meaningful reflection.

A - clear summary of calculations.

A - concise, no time wasted in calculations.

A - linked to aims.

The descriptive statistics were calculated using the data values from Table 2.0. All values were calculated on my GDC.

Table 2.1 showing my descriptive statistics (seconds)

Variables	Run times (x)	Triathlon run times (y)
Minimum Value	1105	1089
Q_1	1139	1206
Median	1180	1251
Q_3	1245	1302
Maximum Value	1301	1404
IQR	106	96
Mean	1189.86	1254.77
Standard Deviation (σ)	60.48	72.69
Lower Limit	980	1062
Upper Limit	1404	1446

The comparison of these two tables allows me to investigate the difference between my run times and my triathlon run times.

Mean

The mean of the two respected time variables are measures of central tendency. The mean is representative of the average value of each distribution. The mean run time is 1189.86 seconds while the mean time for a triathlon run is 1254.77 seconds, both rounded to 2 decimal places. Therefore, the mean run time is significantly lower in a running race, suggesting that on average, I run quicker in a 5km run race to running a 5km run in a triathlon.

D - meaningful reflection.

> **Reflection:**
> Initially, I believed this would be the case. During a running race, fatigue is less prominent and therefore, one would expect the average speed to be quicker.

Standard Deviation & IQR

My data informed that the standard deviation is a wider measure of spread to the mean; it represents the spread of the data around the mean. Therefore, the standard deviation is a great indication to see what data is more consistent with their respected times. The standard deviation for the run time is $\sigma = 60.48$ and for the triathlon run time is $\sigma = 72.69$. This suggests that the run time in a running race has a lower variability and the data is more closely cluttered around the mean, implying the performance in a running race is more consistent.

D/E - meaningful reflection with good knowledge demonstrated.

Interestingly, the IQR for the triathlon run time is lower at 96 seconds compared to a run time in a running race being 106. Unlike the standard deviation where 68% of all data values lie within 1 standard deviation, the IQR shows the spread of the middle 50% of the data. This suggests, that although the time taken in a

D - critical reflection of findings in relation to aims. Student demonstrates thorough understanding of findings.

running race is more consistent when taking in account 68% of the values, when looking at the middle 50%, the running time in a triathlon have a lower variability. This implies that triathlon run times seem to be more consistent when looking at the middle half of the data. However, with the standard deviation being greater, the values outside 50% of the data collection differ more drastically and are more unpredictable to times in a running race. Therefore, although the IQR is smaller for triathlon running times, as shown visually by the box plot, 75% of run times in a 5km running race are faster than the median of the triathlon 5km running times. A greater IQR and therefore, a larger run variability in running races, may be because I have improved more in a 5km race compared to a 5km in a triathlon.

D - excellent reflection.

Reflection:
This information provides an insight into the differences between run time in a running race and run time in a triathlon. Although when looking at the raw data they seem varied, I will now create a box plot to see whether the values are significantly different. Visually representing the data will put the values into context to enable this comparison. *C - thinking creatively to compare results.*

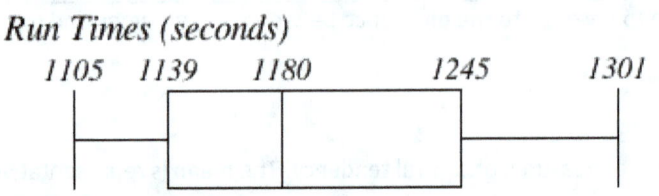

B - multiple forms of mathematical representation in IA: formulae, tables, graphs.

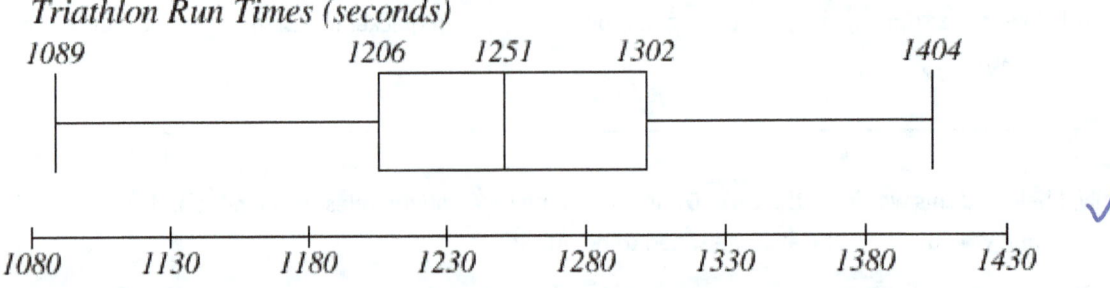

A/B - clear and concise graphs that are simple and easy to read.

Figure 2.2. A box plot constructed with data from Table 2.1 to compare 5km run times in a running race and 5km run times in a triathlon.

Reflection:
I have decided to place the descriptive statistics for 5km run times and triathlon 5km run times on the same axis, as the value do not change drastically and is therefore readable on the same axis. This allows me to visually see the different dispersions of the data sets, enabling a comparison.

The run times are shown to be prominently positively skewed while the triathlon run times are only slightly positively skewed. The median shows that the slowest 25% of the data values for the 5km run times account for more than 50% of the data values for 5km triathlon run times. This suggests that over 75% of data values

D - critical reflection shown here, student has considered possible reasons for findings.

for the 5km run time are quicker than the median value, 1251 seconds, of the triathlon 5km run times. Moreover, there is a clear difference in variability of the data sets. Although the IQR is smaller for the triathlon run times, the data values are less consistent, confirmed by the significantly longer box plot and the standard deviation calculated above. I believe that these descriptive calculations highlight that fatigue significantly affects performance levels. Primarily shown by the greater variability of the triathlon 5km run times, run times in a 5km running race have a lower rate of dispersion and are therefore, more predictable. At the final stages of a triathlon motivation and fatigue can differ, highlighted by the statistical data.

> **Reflection:**
> Interestingly, the fastest run time recorded is in a triathlon, shown by the smaller minimum value. As running is a subset of triathlon, one would expect the time in purely a running race to be quicker as fatigue is less prevalent. However, I have raced in international triathlon races; these are more competitive than local running races suggesting my motivation to perform well was greater. This resulted in my run being quicker.

C - the student has considered the impact of external impacts on their results.

T - Test

A - logically ordered mathematics which supports the development of the IA.

To confirm that my run times in a running race are significantly different to my run times in a triathlon, I will carry out a one tailed $t-test$. As I only want to see whether the run times in a running race are significantly quicker, this form of test fitted the best. I have chosen to perform the test at a 5% significance value as is the most commonly used significance level for t-tests.

Hypothesis:

C - students thoughts and voice can be heard here, they are asking questions of the mathematics and want to explore their investigation further.

$$H_0 \text{ (null hypothesis)} : \mu_1 = \mu_2$$

$$H_1 \text{ (alternate hypothesis)} : \mu_1 < \mu_2$$

B - variables clearly communicated.

Key :

$\mu_1 = mean\ run\ times$

$\mu_2 = mean\ triathlon\ run\ times$

$p\ value = 0.00155$

Figure 2.3 – T-Test calculations of GDC

	A x	B y	C	D
				=tTest_2'
1	1105	1180	Title	2-Samp..
2	1140	1206	Alternate...	μ1 < μ2
3	1146	1212	t	-3.14562
4	1110	1170	PVal	0.001548
5	1139	1089	df	40.6562

D4 =0.001547983987811

E - good knowledge and understanding demonstrated

If:

$p\ value\ <\ significance\ level\ (0.05) : H_0\ is\ rejected\ and\ H_1\ is\ accepted$

$p\ value\ >\ significance\ level\ (0.05) : H_0\ is\ accepted\ and\ H_1\ is\ rejected$

D - meaningful reflection linked to aim 2.

The $p\ value$ is 0.0015. This suggests we will reject the null hypothesis and conclude there is a significant statistical difference between 5km running race times and triathlon 5km run times.

Now, I can confidently conclude that run times in a 5km running race are significantly faster to 5km run times in a triathlon. Although surprising data comparisons, the lowest minimum value of a 5km run time being in a triathlon was because my performance influenced by significant amounts of motivation and international pressure – it is clear that it was a one off performance. The box plot confirms that almost 75% of all 5km run values are quicker than 5km triathlon run values. This highlights the reliability of the $t-test$ and verifies that on average, run times in a 5km running race are quicker and significantly statistically different to triathlon, 5km run times.

D - substantial evidence of meaningful reflections throughout the exploration.

C - students voice can be heard and is invested in their findings.

Conclusion

When investigating the correlation between days since starting training and total time taken in a triathlon, I was expecting a strong negative correlation. The r value reflected my initial thoughts and confirmed my significant progression; I got older and gained more experience my total time taken decreased. The r^2 value reinforced the strong relationship, highlighting the very high level of influence between the two variables. However, after plotting a linear regression line, I believed it hindered the reliability of my claims. When looking at the number of days needing to train for to hit a world record time, the amount of days the graph suggested seemed to be too achievable. Therefore, I decided an exponential model demonstrated my progress more accurately due to the very nature of it. Sports performance is a complex mixture of *"biomechanical function, emotional factors, and training techniques"* (Science, 2023). Performing optimally indefinitely is unlikely to happen, thus the exponential model proved a better fit. However, due to the nature of performance, I believe all models have their limitations. Therefore, despite attempting two different models, a combination of all are necessary to depict performance accurately. Perhaps, the r values and models must be used theoretically rather than conclusively. Moving onto my Second aim, fatigue is shown to clearly affect performance levels. To whether this is to a greater or lesser extent was initially unclear due to the variability of the triathlon 5km run time's data. However, completing a $t-test$ confirmed that run times in a running race are significantly quicker to run times in a triathlon. Through further investigation, I could see if VO_2 max or environmental factors play a part in fatigue levels. This would enable me to more greatly understand my fluctuation in triathlon, 5km run times.

A - IA meets the initial aims and offers a concise summary of findings.

D - critical reflection considering implications upon investigation.

Bibliography

Science, W. o. (2023, September 18). *Sports Performance*. From Encyclopedia.com: https://www.encyclopedia.com/sports/sports-fitness-recreation-and-leisure-magazines/sport-performance

Triathlon, W. (2019, March 8). *Results: 2019 Daman World Triathlon Abu Dhabi | Elite Women*. From https://triathlon.org/results/result/2019_itu_world_triathlon_abu_dhabi/335270

Triathlon, W. (2022, June 11). *Results: 2022 World Triathlon Championship Series Leeds | Elite Women*. From https://triathlon.org/results/result/2022_world_triathlon_championship_series_leeds/545473

Appendix

Triathlon	Distance	Swim Time (s) (750m)	Bike Time (s) (20km)	Run Time (s) (5km)	Date
Grit + Tonic Relay	Sprint	596	1980	1180	16/04/2023
Grit + Tonic	Sprint	621	2010	1206	12/03/2023
JLL Race 2	Sprint	617	2042	1212	04/02/2023
RAK Tri	Sprint	608	2116	1170	15/01/2023
World Champs	Sprint	599	2096	1089	24/11/2022
Fujairah Tri	Sprint	634	2102	1190	17/11/2022
Grit + Tonic	Sprint	660	2099	1214	23/10/2022
Woodhorn	Sprint	587	2351	1221	09/07/2022
JLL Race 2	Sprint	756	2042	1291	17/04/2022
RAK Tri	Sprint	632	2134	1252	20/02/2022
Grit + Tonic Relay	Sprint	662	2088	1200	07/02/2022
WTS	Sprint	619	2180	1265	29/01/2022
JLL Race 1	Sprint	695	2103	1268	17/01/2022
Grit + Tonic	Sprint	628	2159	1278	22/10/2021
Tri Yas	Sprint	744	2174	1228	15/10/2021
Grit + Tonic	Sprint	735	2277	1354	09/04/2021
JLL Race 1	Sprint	757	2342	1302	16/01/2021
Hatta	Sprint	856	2508	1357	13/11/2020
Grit + Tonic	Sprint	642	2324	1404	23/10/2020
RAK Tri	Sprint	760	2478	1250	11/02/2020
Tri Fest	Sprint	778	2412	1354	17/01/2020
Grit + Tonic	Sprint	800	2556	1320	10/01/2020

World Champs	Sprint	599	2096	1089	24/11/2022
Fujairah Tri	Sprint	634	2102	1190	17/11/2022
Grit + Tonic	Sprint	660	2099	1214	23/10/2022
Woodhorn	Sprint	587	2351	1221	09/07/2022
JLL Race 2	Sprint	756	2042	1291	17/04/2022
RAK Tri	Sprint	632	2134	1252	20/02/2022
Grit + Tonic Relay	Sprint	662	2088	1200	07/02/2022
WTS	Sprint	619	2180	1265	29/01/2022
JLL Race 1	Sprint	695	2103	1268	17/01/2022
Grit + Tonic	Sprint	628	2159	1278	22/10/2021
Tri Yas	Sprint	744	2174	1228	15/10/2021
Grit + Tonic	Sprint	735	2277	1354	09/04/2021
JLL Race 1	Sprint	757	2342	1302	16/01/2021
Hatta	Sprint	856	2508	1357	13/11/2020
Grit + Tonic	Sprint	642	2324	1404	23/10/2020
RAK Tri	Sprint	760	2478	1250	11/02/2020
Tri Fest	Sprint	778	2412	1354	17/01/2020
Grit + Tonic	Sprint	800	2556	1320	10/01/2020

IA Sample 10 Examiner Comments: Angry Birds

This is a high quality and creative IA where the student uses the equation of a tangent to determine when to activate the yellow bird's power to hit a pig.

Appropriate for: AASL, AISL

Possible for AAHL, AIHL (For HL you would need more sophistication)

Criterion	SL	HL
A	3	3
B	3	3
C	3	3
D	3	3
E	6	4
Total	18	16

A: It is a nice easy read for a peer (coherent). The structure is generally logical with a clear aim and progression from the problem to the mathematics. Diagrams (e.g. bird's path, position of the pig, tangent point) support the explanation and help understanding. There is some minor inconsistency in formatting or layout that affects flow slightly. Double spacing should have been used.

B: Mathematical notation is generally correct but sometimes equation editor has not been used and this is evident. Variables are defined. Rounding is mentioned and appropriate. Some of the screenshots from GeoGebra are not appropriate as they include more information than needed. This is a main reason for 3 and not 4.

C: The topic shows strong personal engagement. It's clearly the student's own idea and it hasn't been taken from a textbook. The mathematics guides the exploration and the decisions about model and flight clearly come from the student's engagement with their IA.

D: There is a lot of reflection within the IA. The reflection on results helps the student decide what to do next. There are lots of examples shown on the paper. Excellent example on the bottom of page 12.

E: The student demonstrates a thorough understanding of the mathematics which is at an appropriate level for SL. The mathematics is purposeful, correct, and relevant throughout.

IB Mathematics Standard Level Internal Assessment

Title is too long. It should not be just your aim

Determining the optimum launch angle and the optimum position to activate the yellow angry bird's ability, as to hit the green pig on level 17 of Angry Birds 2.

Pages: 12

Table of Contents

INTRODUCTION 3

AIM 3
RATIONALE 3
PLAN 3

DATA COLLECTION 5

PART A: MEASUREMENTS OF LEVEL 5
PART B: FIND THE FUNCTION (QUADRATIC) 5
PART C: FIND THE ANGLE (DIFFERENTIATE) 8
PART D: FINDING THE ABILITY ACTIVATION POSITION (TANGENT) 10

CONCLUSION 13

Introduction

Aim

This exploration aims to find the optimum launch angle for the parabola-shaped projection of the yellow angry bird and therefore calculate the optimum position along the parabola path to activate the yellow angry bird's straight projectile ability, as to hit the green pig on the right on level 17 of the Angry Birds 2 game.

Nice intro. clear aim.

Rationale

As a child, I'd always enjoyed playing Angry Birds. I'd enjoyed trying to get three stars on a level or getting a new high score by taking out the pigs with as few birds as possible. Back then, I mindlessly aimed the bird and hoped it would hit a pig or knock over a tower but as I grew older, I noticed how I could use the curve of the bird's projection, to get over towers or compensate for the curve by aiming higher or lower. Within my IB mathematics course in MYP, when I learned about quadratics and parabolas, it instantly reminded me of the game Angry Birds, with my teacher even using a parabola to model the launch of the bird for an example. I found it fascinating as I never knew this as a child and through the use of math, it opened my eyes to new possibilities and ideas. It had inspired me to get back into playing the game and appreciate the mathematics behind the game. Over the past few weeks leading up to this IA, I had struggled with one level (level 17), in which I couldn't quite hit the pig in the back of the map. I was inspired to conquer this level using mathematics and let me progress.

Within the game Angry Birds, the main objective is to launch a bird using a launcher on the left-hand side of the level and hit the pigs on the right-hand side of the screen. By pulling back on the bird, you can let it go, letting the bird fly. You can control the angle in which the bird is launched, which will affect its height, distance, and direction. Each bird has a special ability that when activated, gives you an advantage when trying to hit a pig or knock over a building. Within my IA, I will use the yellow, triangle-shaped angry bird, which travels in the same parabola-shaped path as the other birds, but when his ability is activated, he travels at a quicker pace in a straight line. This ability can be compared to drawing a tangent on a parabola.

Plan

Before beginning, I had to set up certain tasks before I could get my data to ensure accurate and efficient collection. First, I determined an ideal level in Angry Birds 2. My criteria for that level was that it must include a tower the bird must fly over in the center of the map (ideally the tower's top is at a higher point than the bird in the launcher) and the yellow bird must not be able to hit the pig, based off its projection alone, without using its ability, meaning the pig must be rather far to the right, away from the bird before launch.

Once I had a level, I gathered measurements of the bird's height from the ground, the distance between the pig and the bird (and the midpoint) and the height of the tower in the center, by assuming the launcher is 1 meter and using the launcher as a 'ruler' to make these measurements. I had chosen the launcher to be 1-meter as I had also assumed the increments on the x-axis and y-axis of GeoGebra were one meter, giving me some sort of scale for the level. Finally, I inserted the original screenshot into GeoGebra and using the height of the bird from the ground, line the height (in meters), with the height of the y-axis (1 meter in the game, is one increment on the y and x -axis)

Next, I needed to model the bird's parabola-shaped projection by getting a function that would allow me to fulfill my aim. Using GeoGebra I inserted the photo of my level and the function, $f(x) = a(x - h)^2 + k$, and using GeoGebra's sliders, I manipulated each variable (a, h & k), until a parabola that connects the bird in the launcher and goes over the tower in the middle of the level, but not hitting the pig (line continues to plateau downwards before the pig/ to the left), is formed. Once I found this parabola, I made a note of its function in the form $f(x) = a(x - h)^2 + k$.

After, I needed to calculate the launch angle, that would give me my previously modeled function when playing the level. First, I converted the vertex form of the parabola, into standard form, making the function easier to differentiate. Then, I differentiated the standard form function and after, made $x = 0$. I then took the solution and interested it into $\tan^{-1}$, giving me the angle.

Finally, I needed to calculate the optimum position of the bird, to activate the bird's straight projectile ability. I used GeoGebra to determine the coordinates of the pig, using a screenshot of the level. Next, I inputted the x and y values of the pig's coordinates into the point-slope formula (x_1 and y_1 in $y - y_1 = m(x - x_1)$) and found the possible values of x for the unknown point on the parabola (that will connect to the pig's coordinates). After, I found the value of y by inputting the values of x into the original quadratic in standard form and with the two coordinates for the unknown point, I graphed them on GeoGebra, connected them to the pig's coordinates and determined which set of coordinates would be a better fit (one will not lie on the screenshot and one will).

Data Collection

Part A: Measurements of Level

Before attempting to find a function of a possible parabola, I needed to get measurements of the level. This was to get a scale when plotting the level on GeoGebra with regards to the x-axis and y-axis. Having a scale would allow me to line up with the increments of an x-axis and y-axis on GeoGebra. Furthermore, I used the height of the launcher as a benchmark as it was a standard size across all levels and could give me a consistent measurement. Assuming it was 1 meter in height (Assuming the height to be 1 meter as to match up with the increments of 1 on the x and y-axis of GeoGebra), I used it as a digital ruler to measure the distance away from the pig, the height the bird is from the ground, etc. This was especially useful as it

Figure 1

allowed me to roughly hypothesis where a midpoint of a parabola might lie when graphing on GeoGebra. Overall, one limitation was that the measurements may not be exact due to the fact that the game developers have not defined the height of the launcher, thus by assuming it was one meter, the measurements I made with it may not be true in reality, skewing my findings. As a result of this stage, in GeoGebra, my x-axis and y-axis are labeled in meters, keeping my increments and units consistent throughout the IA.

Part B: Find the Function (Quadratic)

Using GeoGebra, I will get a function in the given form of $f(x) = a(x - h)^2 + k$, the vertex form for quadratics. The value a manipulates the vertical and horizontal strength of the parabola along with if it is a positive parabola or a negative parabola. Manipulating h would translate the function horizontally to the left or horizontally to the right. Finally, k would allow me to vertically shift the parabola vertically upwards or downwards. I used these variables as to model a parabola which would achieve my aim of getting over the center tower without hitting the target pig on its own. Also, most of the numbers within my test functions contained at least two significant figures and at least one decimal place. This was

B+ C+ D+ E+ A good example of how all the criteria are connected

because this was the degree of accuracy presented to me as I created these functions in GeoGebra. Below are my tests and final function I have discovered by manipulating the variables in the vertex form of a quadratic.

The below image should be bigger and clearer

$$f(x) = -0.8(x - 5.4)^2 + 9.2$$

[Test Function 1] Unfortunately, the bird would not go over the middle tower before hitting the pig and therefore would not satisfy my aim. I had chosen this as my initial function because I was mainly looking for a function which would have a vertex, which is higher than the middle tower. With that value of k in mind, future functions should have a k-value around $k \geq 9$ as to ensure the bird goes over the tower. Although the value of k (vertical translation) gives a vertex which is higher than the tower, the parabola still isn't wide enough to give the bird a range which allows it to go over the tower and towards the pig. If I don't want to knock over the tower, I need to manipulate the value of a to extend the range.

Figure 2

$$f(x) = -0.1(x - 6.6)^2 + 7.6$$

[Test Function 2] The projection does have a decent range which travels past the tower, but due to the stretch of the parabola, I also had to alter the value of h and k in order for the parabola to connect the bird in the launcher. Next I was testing to see what the minimum value would of a have to have a function which is stretched enough to get to the other side of the middle tower. As a result of manipulating these variables, despite the range being long enough to reach the other side, the bird still does not fly over the tower without hitting it and knocking it over. The stretch of the parabola is a step in the right direction as the range is great enough, but the bird will not fly over the tower in order to get to the other side (All my specific values from test 1 had changed in test 2. The reason why is explained below).

Figure 3

$$f(x) = -0.1(x - 10.4)^2 + 11.4$$

Figure 4

[Test Function 3] Although the bird gets over the tower without hitting it, it flies over the target and the path taken seems to be unrealistic with regards to the game itself, due to the launcher not being able to launch the bird that high nor that far. I had gotten to this function because I was manipulating both the h and k values. Now, I need to reduce the value of k as to lower the height of the parabola, thus the bird would be less likely to fly over the target (yet still manage to get over the tower). As a result, I will have to slightly reduce the values of a and h to compensate adjusting k, which would disconnect the parabola from the bird (The parabola must light up with the bird in the launcher as the launcher is the bird's starting point of the parabola-shaped projection).

Final Function   D+ Lots of good reflection

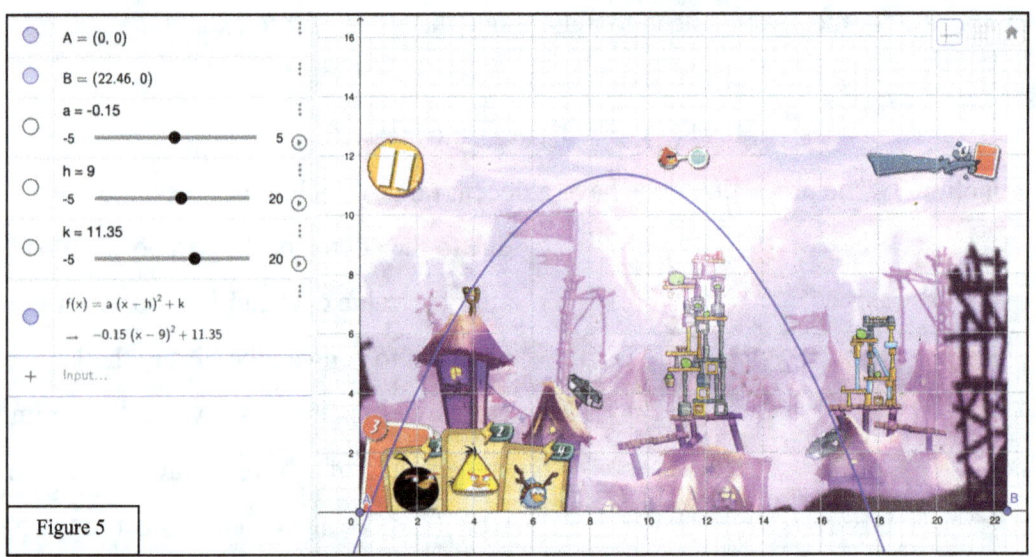

Figure 5

$$f(x) = -0.15(x - 9)^2 + 11.35$$

I had decided upon this function, mainly because it allowed the bird to pass over the tower in the middle, without hitting it, and because it would allow the bird to reach closer to the bird's maximum distance. This was crucial because the parabola allows the bird to just pass over the top right corner of the tower, narrowly avoiding the small barrel, which if hit, would alter the bird's flight path. Furthermore, the parabola does not directly connect the pig and the bird, meaning the ability is required to hit it (The bird cannot hit the pig without the ability in the first place). With regards to my third test, after reducing the value of k, it allowed the bird to still get over the tower, but once it got over, it would not fly over the pig. Throughout my tests, modifying one variable had forced me to change another. This was because if I were to increase the value of a, although the vertex would remain in the spot (because k had not changed), it would still need to be altered because the parabola no longer is connected to the bird (which it needs to be if the parabola models the flight path of the launch) and so the value k must be altered as to lower the parabola and connect it to the bird in the launcher. Back to referring to my third function, I slightly modified the values of a and h as to allow the parabola to connect with the launcher once again. The result was a trajectory that allowed the bird to fly over the tower without hitting it, not flying over the pig and a sketched parabola which still was connected to the launcher. Reflecting on this stage, one limitation in my IA is the fact that it is very difficult to accurately test the function as I can't input a function or numbers and see how it would affect the bird. Although I still had done a rough simulation of my possible parabola in the game as to see if the trajectory simulated by my parabola was similar. I had found that like my modeled parabola, I was able to launch the bird in a certain way which allowed the bird to get over the tower, narrowly avoiding the barrel on the tower.

Part C: Find the Angle (Differentiate)

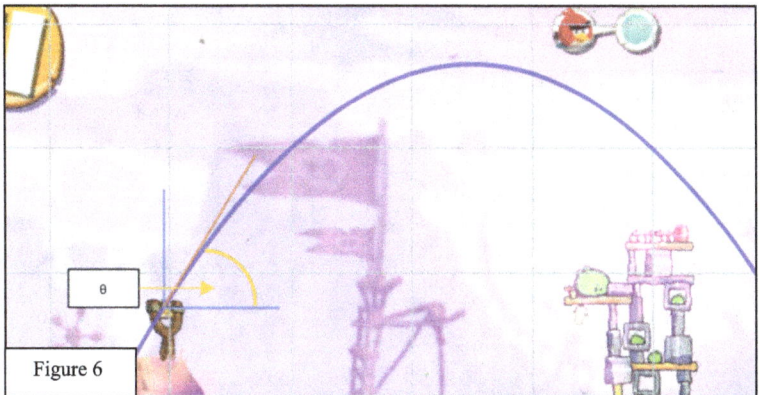

Figure 6

I had first converted my function from vertex form to standard form due to standard form making the next step, differentiation, a bit easier to follow due to the lack of brackets, avoiding potential mistakes.

$$f(x) = -0.15(x-9)^2 + 11.35$$
$$f(x) = -0.15(x^2 - 18x + 81) + 11.5$$

$$f(x) = 0.15x^2 + 2.7x - 12.5 + 11.35$$
$$f(x) = -0.15x^2 + 2.7x - 0.8$$

After, I took the first derivative of my function in standard form. This was used to find the gradient of the parabola I had modeled.

$$f'(x) = -0.3x + 2.7$$

Next, I had substituted x to equal 0, giving me a value of the tangent of the angle which I can use to find the launch angle.

$$f'(x) = -0.3x + 2.7$$
$$f'(0) = -0.3(0) + 2.7$$
$$f'(0) = 2.7 = \tan(\theta)$$

I finally found the inverse of tan as to give me the angle I should launch the bird at. This is the angle required to launch the bird as to create the optimum parabola with the function $f(x) = -0.15(x - 9)^2 + 11.35$. Here I rounded to 2 decimal places due to the fact that in the next step, I defined my coordinates of the pig to two decimal places (as provided directly from GeoGebra) and I decided to keep my data to two decimal places to keep my rounding consistent. To conclude this section, a limitation I found was that I am unable to properly test the use of this angle in the situation of the game. I cannot input a certain angle, but rather the game does not tell me what angle I am launching at and I need to do it by eye. This makes it more difficult to verify this angle and use the specific data to complete the level.

$$\tan^{-1}(2.7) = \mathbf{69.68°}$$

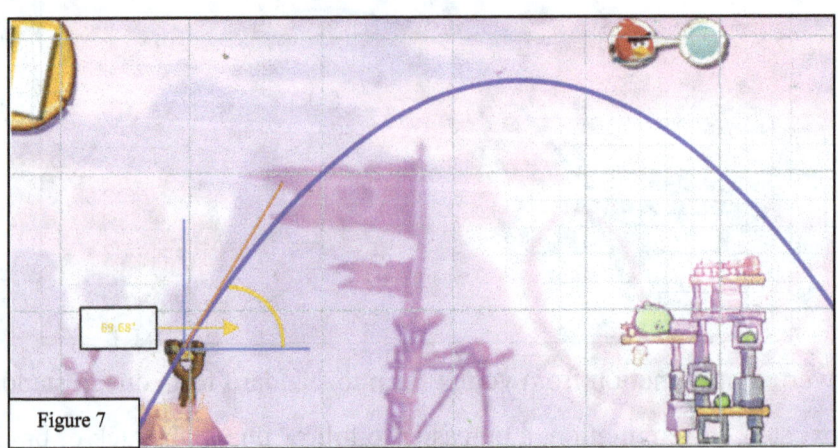

Figure 7

Part D: Finding the Ability Activation Position (Tangent)

One thing to note, is that the number of decimal places is not consistent in some areas of the calculations. This is because all the values are taken directly from GeoGebra, which gave all my values to a certain number of decimal places and so I tried to use as many decimal places as possible in my calculations as to be as accurate with my numbers as possible. Whenever I made a calculation, I tried to have it at three decimal places as to increase accuracy a little bit more. According to GeoGebra, the center of the pig's body was roughly around the coordinates, (17.45, 5.95) on the x-axis and y-axis. I chose these points due to the fact that it is very close to the center of the pig's body (I did have to make this assumption and so the pig's real center might be different, giving me a different set of coordinates to input into my point-slope formula. This is a data limitation), allowing me to hit the pig directly on with a greater amount of accuracy. The coordinates I should launch the bird at as it follows my found parabola are unknown and what I am now looking for:

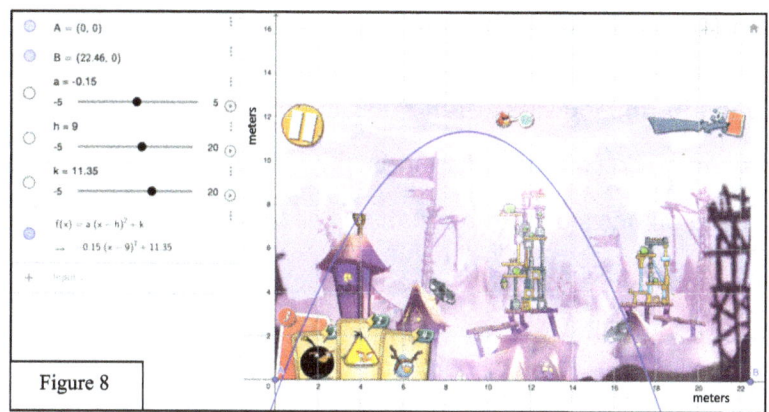

Figure 8

I let a be the x coordinate of where I'd activate the ability. I defined it as a rather than x due to the fact that I already have a defined value of x within the original formula.

$$y - y_1 = m(x - x_1)$$

$x_1 = a$, my unknown x-coordinate for the tangent: a

$y_1 = $ The original function of the parabola in standard form: $f(a) = -0.15a^2 + 2.7a - 0.8)$

$y = $ The y-coordinate of the pig: (5.95)

$x = $ the x-coordinate of the pig: (17.45)

$m = $ The first derivative of my original function: $f'(a) = -0.3a + 2.7$

My first step was to insert my function of the parabola in standard form, into the value of y_1 and my differentiated function in m, the gradient.

$$y - (-0.15a^2 + 2.7a - 0.8) = (-0.3a + 2.7)(x - a)$$

Page of

E+ This whole section demonstrates an excellent understanding of a difficult concept.

Next, I substituted the coordinates of the pig in y and x_1.

$$5.95 - (-0.15a^2 + 2.7a - 0.8) = (-0.3a + 2.7)(17.45 - a)$$

In order to find the value of a, I needed to expand my equation as to remove the brackets, allowing me to combine like terms, then get a quadratic function for a. I also needed to get all terms on one side of the equal sign as to make a function that equals zero, allowing me to finally determine a value for a.

$$5.95 + 0.15a^2 - 2.7a + 0.8 = -5.235a + 0.3a^2 + 47.115 - 2.7a$$
$$0.15a^2 - 2.7a + 0.8 = 0.3a^2 - 7.935a + 41.165$$
$$-0.15a^2 + 5.235a - 40.365 = 0$$
$$0.15a^2 - 5.235a + 40.365 = 0$$

After getting my function, I needed to find my two values of a, which would give me two possible x values for the position along the parabola to activate the bird's ability.

$$a = \frac{5.235 \pm \sqrt{(-5.235)^2 - 4(0.15)(40.365)}}{2(0.15)}$$

$$a = 23.4 \text{ or } 11.5$$

I used my two values of x to find two possible values for y, giving me two sets of coordinates where I might be able to launch the bird's ability by inserting them into the original function in standard form.

$$f(23.4) = -0.15(23.4)^2 + 2.7(23.4) - 0.8 = -19.754$$
$$f(11.5) = -0.15(11.5)^2 + 2.7(11.5) - 0.8 = 10.4125$$

$$(23.40, -19.75) \text{ or } (11.50, 10.41)$$

The tangent is connected between the points **(17.45, 5.95) for the pig** and **(11.50, 10.41) for the bird's position** (I did not use the other possible set of coordinates as for reasons stated below in its analysis). I should activate the bird's ability when it is at (11.50, 10.41), as to hit the pig at (17.45, 5.95). Note that throughout the calculations, my final answers were given to two decimal place because the coordinates of the pig from GeoGebra were given to two decimal places and I needed the coordinates to launch to have the same number of decimals as well (so when the two sets of coordinates are compared, they have the same degree of accuracy).

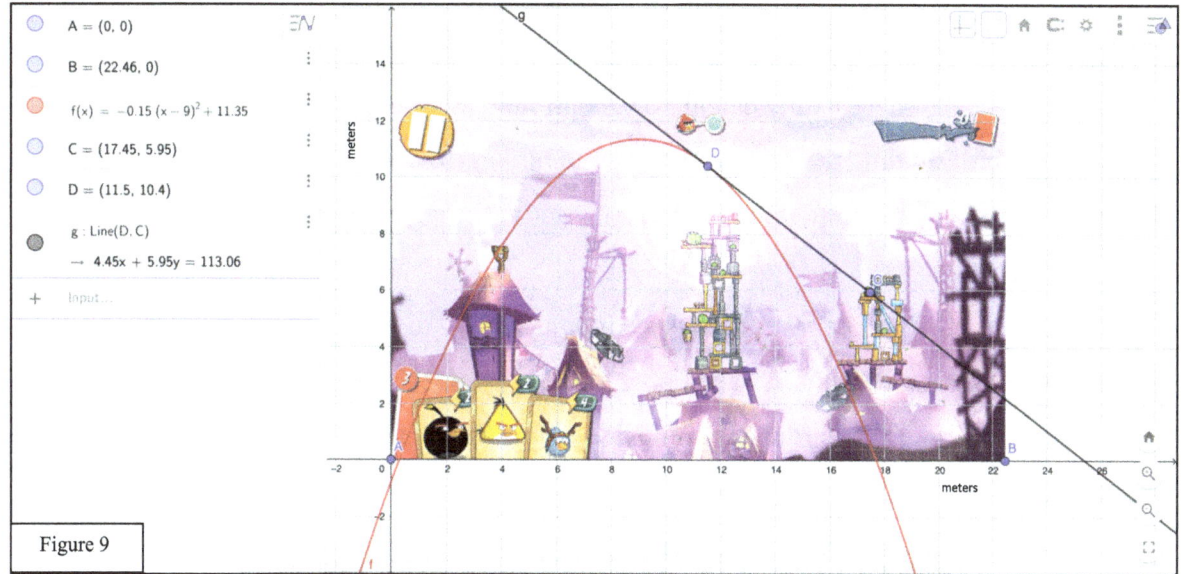

Figure 9

To conclude the data collection, my data informs me that when trying to beat this level, I should launch the yellow angry bird at 69.68° within the launcher, with the intention of creating the parabola-shaped launch of $f(x) = -0.15(x - 9)^2 + 11.35$, which would allow the bird to fly over the center tower, then followed by activating the bird's ability when at coordinates (11.5, 10.4), as to be projected in a straight line (forming a tangent to the parabola at the stated coordinates), therefore hitting the pig directly, at point (17.45, 5.95).

Final Data Summary

Try to be consistent with type of x that is used. Use equation editor.

Launch Angle	69.68°
Function for Parabola	$(x) = -0.15(x - 9)^2 + 11.35$
Pig Coordinates	(17.45, 5.95)
Ability Activation Coordinates	(11.50, 10.40)

All values I <u>personally calculated</u> are given to four significant figures

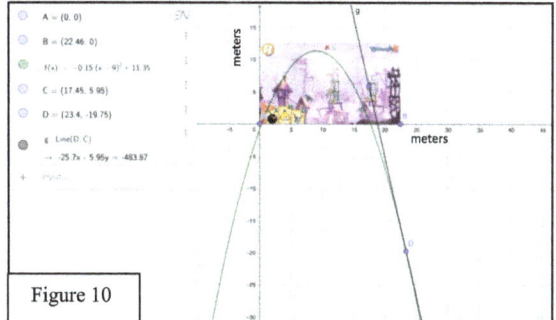

Figure 10

Here is my other possible set of points of the tangent between $(23.40, -19.75)$ and (17.45, 5.95). It does connect the parabola to the pig, but this tangent would not work. This is due to the fact that for one, the point 'D', is off the screenshot/ map, which in the context of the game, would not be possible to launch at that position as the bird is considered 'dead' when it is not visible on the screen. Furthermore, the bird's ability always launches downwards to the right, not upwards to the left as seen in this tangent compared to my used tangent.

Conclusion

I was able to successfully calculate the optimum parabola-shaped trajectory of a yellow angry bird, finding the angle to launch the bird at (to achieve this trajectory) and finally the optimum point along the trajectory to activate the yellow bird's straight-line projectile ability.

Mathematically, I believe there is a strong accuracy to my results due to the use of GeoGebra to give me the numbers and data required. GeoGebra had given me far more accurate numbers for my coordinates compared to creating a grid myself and looking for the coordinates. Furthermore, based on the photo, it allowed me to create a function, in real-time, which I could manipulate with extreme precision (with many decimal places) as to model the exact function I was looking for, making as many adjustments as needed. As a result, I was able to get accurate numbers and data I could use in my formulas and visualize what I was doing to ensure my points or parabola worked for my aim and in the context/physics of the game.

A large limitation to my IA and the data would be the difficulty in testing my angle and ability launching position in the game itself. The purpose of this IA was to find an angle and ability launching position to allow me to hit the target I was aiming for. The game does not allow me to input a specific launch angle and position to activate the ability (instead, I'd have to manipulate these variables by hand, without the game showing me my angle or coordinates I'm launching at in real-time). As a result, I am only left with an estimation based on photos from GeoGebra. I was able to do a few tests on the level, estimating my angle and where I wanted the bird to end up and did manage to either get very close or even hit the pig! Next time, my biggest suggestion would be using a simulation to visually see the bird being launched and possibly hitting the pig.

After completing the IA and using the data I had collected, I had managed to beat this level, after being stuck on it for quite some time. Although I couldn't input the specific data points I found, by using my diagrams and estimating, I managed to get a similarly shaped projection and did manage to hit the pig with my ability on my second try (similar to what is displayed in figure 9).

IA Sample 11 Examiner Comments: Voronoi and TSP

This is an excellent AIHL IA where the student uses both Voronoi Diagrams and Graph Theory to explore university locations/choices.

Appropriate for: AIHL, AISL (SL wouldn't need the graph theory)

Possible for AAHL, AASL (if you are willing to learn new content from the AI course)

Criterion	SL	HL
A	3	3
B	3	3
C	3	3
D	2	2
E	6	5
Total	17	16

A: The exploration has a clear aim and conclusion. It is coherent and usually well organized. Some of the diagrams are not perfectly clear, and the two separate aims makes the IA not concise hence the lost mark. Page numbering missing.

B: The mathematical communication is appropriate and generally consistent. The weighted graph is a screenshot and clarity is lacking. Tables are well done, and variables are defined. Use of equation editor would have helped achieve 4/4.

C: The topic is original and has a clear personal relevance. The decision to combine two distinct areas of mathematics (spatial optimisation and path planning) shows initiative and independent thinking. The student is clearly engaged with the mathematics and the results guide the exploration. This student learned graph theory independently which the teacher would have noted.

D: The student reflects throughout but there was opportunity for more critical reflection and to explore different methods which is why 3/3 was not awarded.

E: The student uses HL maths and demonstrates a good understanding which makes 5/6 possible. All techniques are applied appropriately and correctly for the level of complexity involved. More rigor and sophistication required to achieve 6/6 in HL.

Comparing University Options through Voronoi Diagrams and Graph Theory

Mathematics: Applications and Interpretation HL Internal Assessment

Page count: 22

I. **Introduction**:

Rationale:

As a final-year IB diploma student, I experience a mix of excitement and anxiety about the university application process, which feels like a constant weight on my shoulders. Numerous factors contribute to this pressure, including the need to excel in my IB exams to meet the required grades, preparing for the UCAT, and crafting a compelling personal statement. Another factor to consider when deciding which university, I want to apply to is the first and least pressured hurdle. Of course, this depends on many factors, such as academic requirements and majors, as this will impact whether I become a doctor. The first requirement is the location, as it is a key consideration for me. Since my family will remain overseas, having airports nearby would make it easier for them to visit me on holidays. The second requirement before making a decision about which university to attend, is to visit each campus to familiarize myself with their environments and the facilities on campus. Hence to accomplish such a task I need to travel to each university, by train, which is the best option available compared to taxi, airplane and bus. As a student with a limited budget, my second target is to find the most affordable route for visiting each of these universities with the lowest sum of money spent.

Aim:

A - Two realistic aims with extra details of how they are achievable

The first aim of this IA is to locate the optimal university based on its distance to airports and the number of airports. A Voronoi diagram will be used to analyse the distances between each of the 7 selected universities and the nearby airports. The second aim of this IA is to estimate the costs required for visiting each campus, the **Traveling Salesman Problem** (TSP) problem will be used to find an estimate of the costs. To answer the (TSP) a weighted graph that represents direct train routes will be used to calculate the least-cost Hamiltonian cycle a route that visits each campus once (Wolfram MathWorld). Travelling Salesman Problem (TSP) will give a range of costs by calculating the upper and lower bounds. However, the solution derived will be an estimate of the costs, this is significant because this may mean I would have to carry more money as I am uncertain whether in reality is it lower or higher than the calculated value.

Short-listed universities:

The 7 universities I have selected are Cambridge University, University College London, University of Leeds, University of Manchester, University of Birmingham, University of Newcastle and University of Bristol. One of the reasons these universities were selected is because they offer post-graduate residency programs for surgery with the NHS. More specifically, I am drawn to University College London for its Medicine program because of its emphasis on integrating cutting-edge research with hands-on clinical experience. This particularly interests me as I can use the latest medicine to save patient's lives and watch them recover (hopefully).

C - Has done his research and provided good reasons to choose the universities.

Exploration:

Voronoi Diagram construction:

A Voronoi Diagram is "a type of tessellation pattern in which a number of points scattered on a plane subdivides into exactly n cells enclosing a portion of the plane that is closest to each point." (Bellelli). In constructing the Voronoi diagram, first I overlaid a Google Maps image with Universities in England over a grid using (Geogebra). Figure 1 below shows the GeoGebra graph with a square grid over it with a scale of 25km per 1 unit, with the 7 universities as blue points (CU, UCL, UL, UM, UB, UNC, UBr). The coordinates are recorded in Table 1.

Figure 1: Shows the map of shortlisted universities

Defined and accurate.

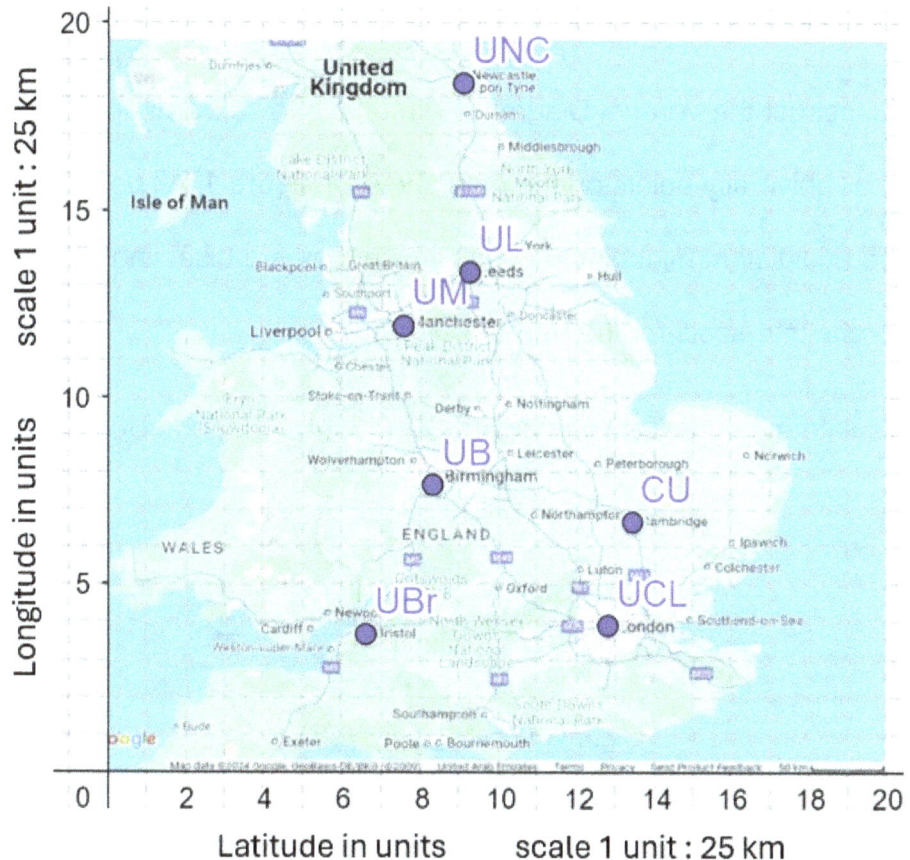

B - Diagram is labelled well with clear locations of universities

Before even thinking about constructing the Voronoi Diagram, the coordinates of university sites/points must be determined so it can be used to calculate the perpendicular bisectors that form the borders of a Voronoi cell. This is shown in Table 1 where the coordinates of each site are in latitudinal and longitudinal units in the form of (latitude, longitude).

B - Good use of table.

Table 1: Shows the coordinates of universities

University Names	Point Name	Coordinates
Cambridge University	CU	(13.44, 6.67)
University College London	UCL	(12.81, 3.91)
University of Leeds	UL	(9.26, 13.34)
University of Manchester	UM	(7.57, 11.90)
University of Birmingham	UB	(8.32, 7.67)
University of Newcastle	UNC	(9.09, 18.39)
University of Bristol	UBr	(6.62, 3.67)

In order to construct the Voronoi Diagrams I first have to calculate the perpendicular bisectors of all university points on the map shown in figure 1. All perpendicular bisectors are equidistant/midpoint to points which they are bisecting. These will be the edges or the intersects for the Voronoi Diagram.

Firstly, I calculate the edge between points CU (13.44, 6.67) and UCL (12.81, 3.91).

$$Midpoint\ of\ [CU\ \&\ UCL] = \left(\frac{x_1+x_2}{2}, \frac{y_1+y_2}{2}\right)$$

$$= \left(\frac{13.44+12.81}{2}, \frac{6.67+3.91}{2}\right)$$

$$= (13.125, 5.29)$$

To find the perpendicular bisector, the gradient of the line between points CU and UCL is substituted into the formula of $-\frac{1}{m}$ which is the negative reciprocal of the gradient m.

$$m \text{ of } [CU \text{ \& } UCL] = \frac{y_2 - y_1}{x_2 - x_1}$$

$$= \frac{3.91 - 6.67}{12.81 - 13.44}$$

$$m \text{ of } [CU \text{ \& } UCL] = \frac{92}{21}$$

$$Gradient \text{ } of \text{ } perpendicular \text{ } bisector \text{ } of \text{ } [CU \text{ \& } UCL] = -\frac{21}{92}$$

The equation of a straight line is $y = mx + c$ next, I substituted the coordinates of the midpoint in x, y and m to find the *y-intercept* (c). This gives me the linear equation of the perpendicular bisector.

$$y = mx + c$$

$$5.29 = -\frac{21}{92} \times 13.125 + c$$

$$c = 5.29 + \frac{2208}{736}$$

$$c = 8.286$$

$$y = -\frac{21}{92}x + 8.286$$

This linear equation shows the edge between CU and UCL in other words the perpendicular bisector between points CU and UCL.

E - Not receiving nor losing points for this section. It looks relevant as it shows some understanding, but at a level of math that is easy for the course.

These calculations were performed for all other edges between points. To enhance efficiency, I used Microsoft Excel to make a spreadsheet to calculate the midpoints and then use that to find gradients of the remaining perpendicular bisectors, utilizing the math tool in Excel. The data was kept unrounded to maintain precision for subsequent calculations.

B - Degree of accuracy considered

Table 2. Shows the results calculated in the spreadsheet

Universities	x1	y1	x2	y2	m	Perpendicular m	midpoint for x	midpoint for y	c
CU & UCL	13.44	6.67	12.81	3.91	4.380952	-0.22826	13.125	5.29	8.286
UCL & UBr	12.81	3.91	6.62	3.67	0.038772	-25.7917	9.715	3.79	254.356
UL & UM	9.26	13.34	7.57	11.9	0.852071	-1.17361	8.415	12.62	22.496
UM & UB	7.57	11.9	8.32	7.67	-5.64	0.177305	7.945	9.785	8.376
UB & CU	8.32	7.67	13.44	6.67	-0.19531	5.12	10.88	7.17	-48.536
UNC & UL	9.09	18.39	9.26	13.34	-29.7059	0.033663	9.175	15.865	15.556
UBr & UB	6.62	3.67	8.32	7.67	2.352941	-0.425	7.47	5.67	8.845
UB & UCL	8.32	7.67	12.81	3.91	-0.83742	1.194149	10.565	5.79	-6.826
UBr & UM	6.62	3.67	7.57	11.9	8.663158	-0.11543	7.095	7.785	8.604
UM & UNC	7.57	11.9	9.09	18.39	4.269737	-0.23421	8.33	15.145	17.096
UL & UB	9.26	13.34	8.32	7.67	6.031915	-0.16578	8.79	10.505	11.962
CU & UL	13.44	6.67	9.26	13.34	-1.59569	0.626687	11.35	10.005	2.892
CU & UM	13.44	6.67	7.57	11.9	-0.89097	1.122371	10.505	9.285	-2.506
UM & UCL	7.57	11.9	12.81	3.91	-1.5248	0.65582	10.19	7.905	1.222

Using the data from Table 2, I constructed the Voronoi Diagram in graphing tool (Geogebra) shown in Figure 2. The edges are the perpendicular bisectors between university sites/points. These edges indicate the set of midpoints universities, but they end when another site/point is closer to the edge than the points it bisects.

Figure 2. Displays the Voronoi diagram of university locations

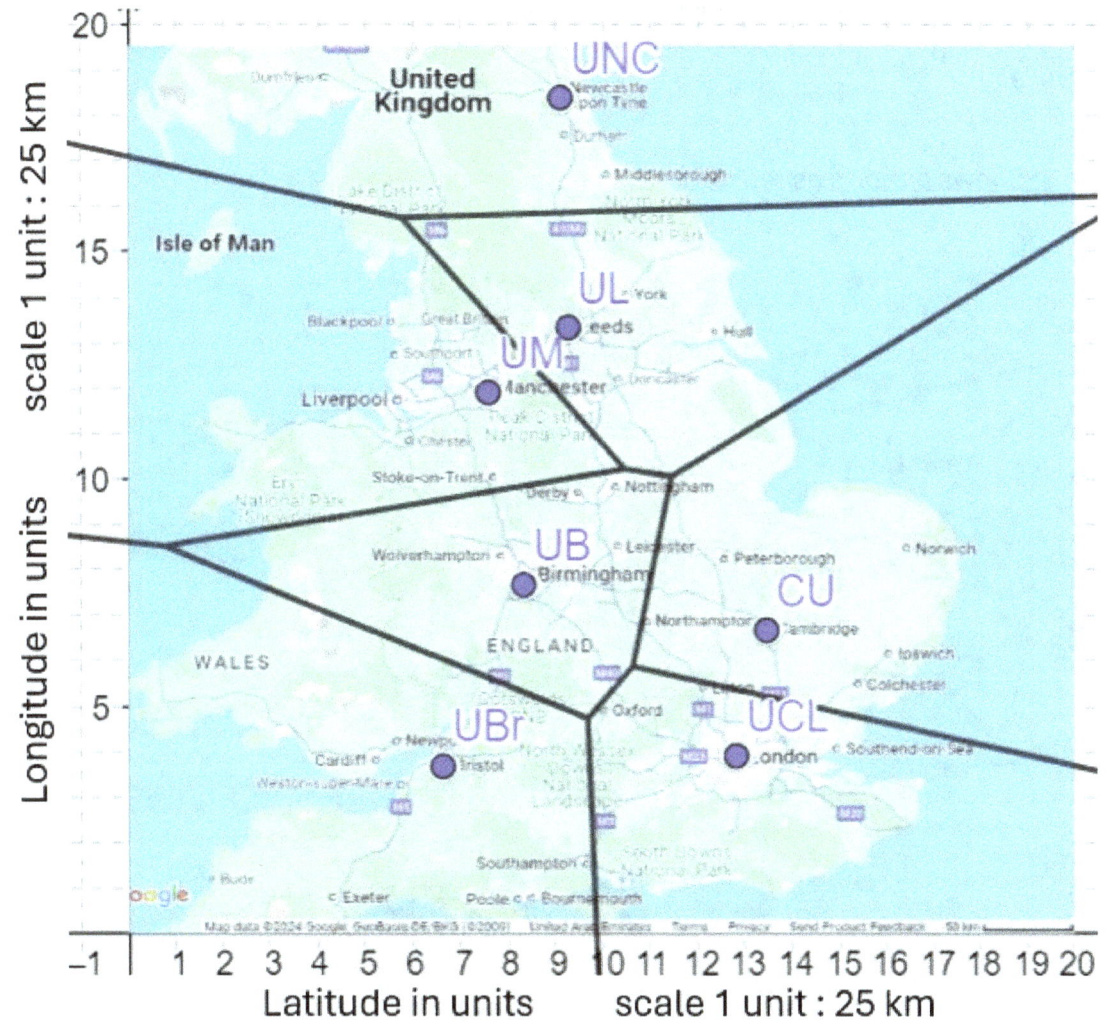

Voronoi Diagram Analysis:

To identify the Voronoi cell with the most optimal location, more specifically the greatest number of nearby airports, I considered airport proximity as a key factor in my university selection, as mentioned in the introduction. By adding airport locations to the grid, we can easily determine which university has the most airports nearby. In Voronoi diagrams, each airport falls into the cell of the university it is closest to, following the general rule that each point within a cell is nearest to its corresponding site *(Bellelli)*.   C - Engagement in the form of further research can be seen throughout through accurate sourcing.

Figure 3: Shows airports as red dots

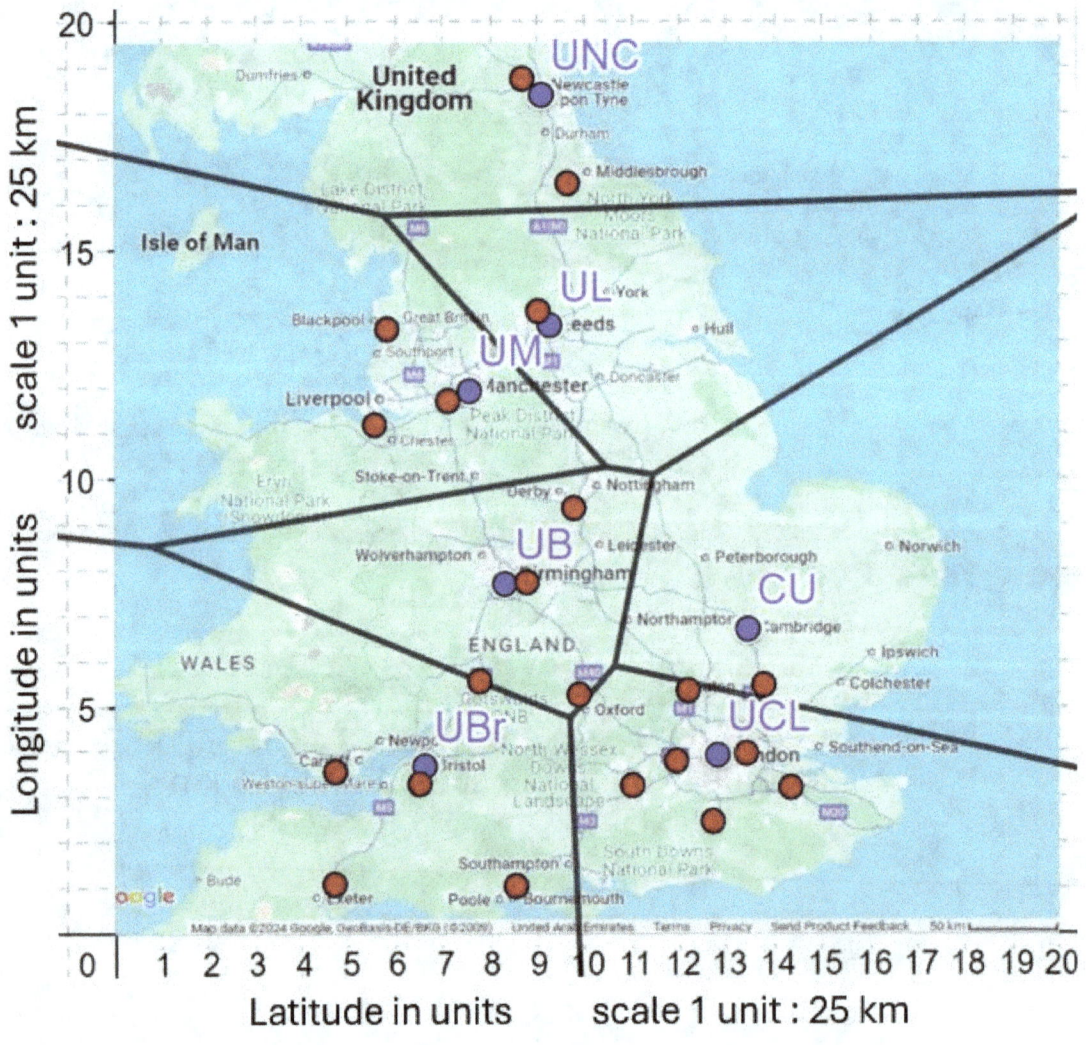

The map above (figure 4) shows the locations of England's international airports on the Voronoi Diagram illustrated by red dots on (Geogebra). Each cell with the red dot(s) indicates the nearest university to that airport. I counted the number of red points in every Voronoi cell, via inspection, and recorded my results in table 3 below. Additionally, for any airports on cell boundaries, I included it in both cells since it is equidistant from the points on the edge to either point. This is a limitation for Voronoi

D - Considers limitations and acts on them by defining what he will do.

Diagrams as being equidistant between 2 points doesn't mean it is the closest, for example, the airport between UB and UBr. Assuming all universities are perfectly latitudinal and longitudinal and are latitudinally closer together while longitudinally far apart, then the Voronoi cell would look like rectangles, and if the airport is located on the longest midpoint of a rectangular cell it will still seem close relative to the cell, while being much further away than the airport located on the shorter midpoint.

Table 3: Displays the number of airports at each point

University point (Voronoi cell)	Number of International Airports
CU	1
UCL	6
UL	1
UM	3
UB	4
UNC	2
UBr	5

From Table 3, we see that cell UCL has the most international airports, with a total of six. This makes University College London the university with the most international airports, positioning it as an ideal location. Surprisingly, Cambridge University and the University of Leeds have fewer airports in their cells on the Voronoi diagram, with counts of 1 and 1, while the University of Bristol has the second most with 5. However, when I was examining the graph, I noticed that although Cambridge

University is near several airports, only one falls within cell CU due to the proximity of UCL and UB sites, which reduced the number of airports near Cambridge.

Therefore, to determine whether the results gathered were affected by the nearby sites. First, the distance between site UBr and the farthest airport located inside UBr's Voronoi cell was calculated. The equation for a circle based on two points was found using (Geogebra) in the following format: $(x - h)^2 + (y - k)^2 = r^2$, where (h, k) is the coordinates of a point or the circle's center (in this case, points UL and CU), and r is the radius. (Geogebra) calculated that the circle's equation around UBr was $(x - 6.62)^2 + (y - 3.67)^2 = 10.61$ I calculated the radius of $10.61 = r^2 \therefore r = 3.67$ units (3 s.f.). Sites UL and CU coordinates are used as the circle center and both have a radius of 3.67 units. Thus, to find the real-life distance of 3.67 times this by 25 km to find the radius in kilometers 3.67 x 25 = 91.75 km which is about an hour's drive at 100kph disregarding traffic and speed limits, which is a reasonable time for me.

E - bringing in circles (which is not on the course). It is not overly complicated, it does add significantly to his decision about his first aim and it is accurate. Sophistication shown as it looks at the problem from a different perspective that helps him make a better decision.

Figure 4: depicts a map of universities and airports with circles UL and CU

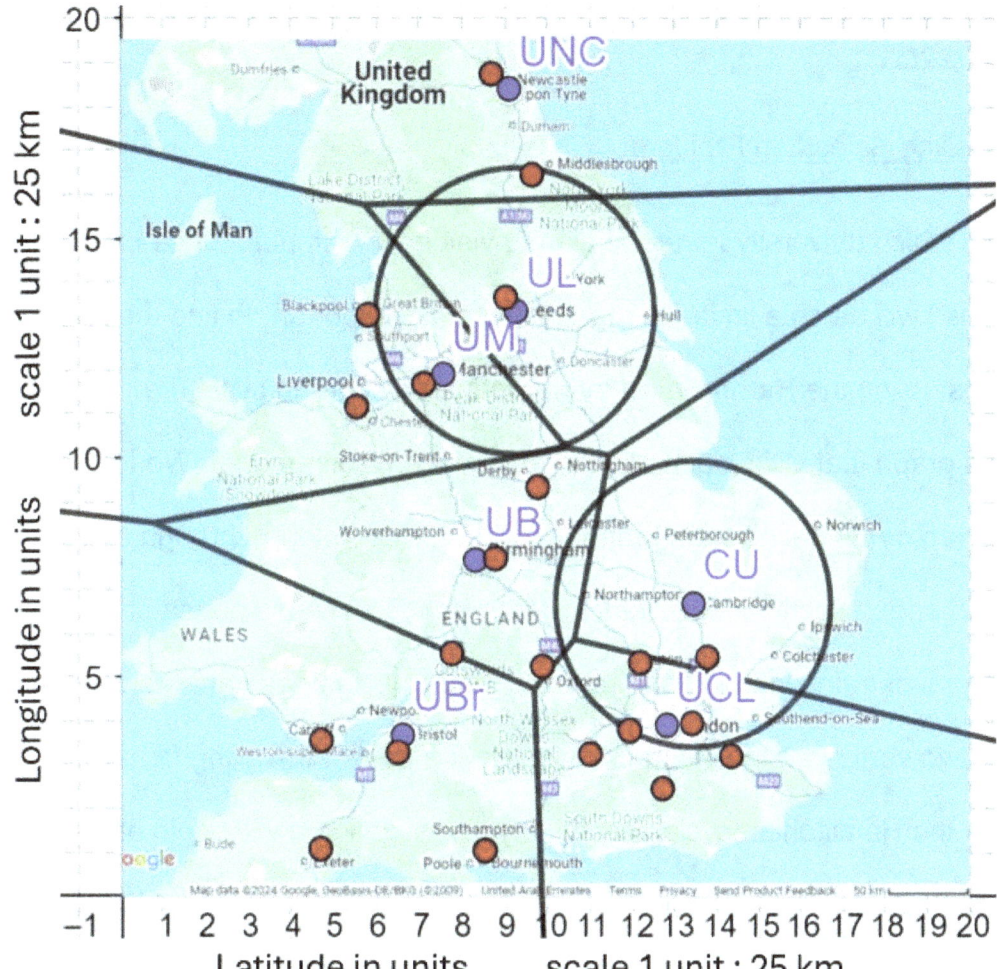

As shown in Figure 4, these circles reveal that 3 airports fall within the area of circle UL, in contrast to only 1 airport within Voronoi cell UL. As a result, this highlights another limitation of using Voronoi diagrams because if the sites/points are closer together compared to spread out it would seem that there are very few airports per university, as the denominator increases because airports are divided/shared by more universities. This can be seen in circle CU than in circle UL as there are more airports in circle CU, despite both circles with the same radii (thus the same areas). This has changed my ranking of the university options as Cambridge University has 3 more airports than before, now combined with its reputation I have selected it as my second choice.

D - Reflection in changing his ranking based on his new improved method. This is critical reflection considering different perspectives and discussing the strengths of it.

Graph Theory:

The Travelling Salesman Problem

To decide which university I prefer, I plan to visit each campus. Since I am a student this means I will be on a limited budget. Hence, it's important to find the lowest weight/cost using the **Hamiltonian cycle** definition: a graph cycle (i.e., closed-loop) through a graph that visits each node exactly once (Wolfram MathWorld), rather than the **Eulerian cycle** definition: a graph cycle (i.e., closed-loop) through a graph that visits each edge exactly once (Wolfram MathWorld), because nodes or vertices are the locations of train stations near each university while edges are the weight/cost between two vertices. Since I am focusing on the cost of travelling from university to university the Hamiltonian cycle best fits my investigation. I will begin and end at UCL as the data from my Voronoi concluded that UCL has the most airports nearby. I will do this by approximating an upper and lower bound of the **practical Traveling Salesman Problem** (TSP) definition: practical TSP requires that each vertex must be visited at least once before returning to the start vertex (Naomi). To identify the least cost train route. A weighted graph was made to solve the TSP, which shows the weighting as train fares in British pounds (£) this is only for direct connections from the stations nearest to each university. In this graph, edge weights represent train costs, while vertices correspond to the various train stations rather than their actual geographic locations. I chose the TSP over the **Chinese Postman Problem** because It is an Eulerian circuit problem that focuses on minimizing the total cost of traversing all edges rather than nodes, with the ability to traverse some edges multiple times, which is not needed in my IA as I will visit each node once.

Figure 5: Shows a weighted graph of direct train journeys   B - Clear and easy to follow.

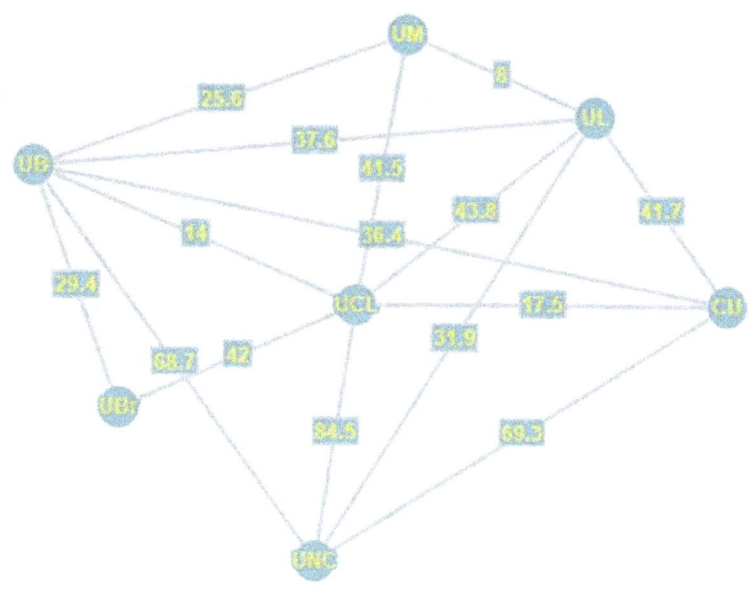

Table 4 below is a simplification of the weighted graph above with the weights of the adjacency of the edges to vertex inputted into the table.

B - Using tables to go with his graphs providing variety of mathematical representation

Table 4: Shows a weighted table of direct train journeys

Vertex	UCL	CU	UL	UM	UB	UNC	UBr
UCL		17.50	43.80	41.50	14.00	84.50	42.00
CU	17.50		41.70	-	36.40	69.30	-
UL	43.80	41.70		8.00	37.60	31.90	-
UM	41.50	-	8.00		25.60	-	-
UB	14.00	36.40	37.60	25.60		68.70	29.40
UNC	84.50	69.30	31.90	-	68.70		-
UBr	42.00	-	-	-	29.40	-	

This '-' symbol means that there is no direct connection between these vertices.

In order to solve the **classical Travelling Salesman Problem** (TSP) definition: **classical** TSP requires that each vertex is visited exactly once before returning to the start vertex (Naomi). I must determine which Hamiltonian cycle in a weighted complete graph has the least weight. The graph must be considered a real TSP if it is not complete or does not satisfy the Euclidean property, which states that the direct path between any two vertices is not always the shortest. I must determine which Hamiltonian cycle in a weighted full graph has the least weight in order to solve the classical Travelling Salesman Problem (TSP). A complete graph is an undirected graph in which every vertex is connected by an edge to every other vertex. As shown in Table 3, some vertices lack direct connections to every other vertex (for instance, vertex UBr is only directly connected to vertices UCL and UB), making Figure 5 an incomplete graph. Therefore, it is not possible to solve it as a classical TSP, and the problem must be approached as a practical TSP instead.

Finding the least weighted path that begins and finishes at the same vertex while making at least one visit to each vertex is the goal of the practical Travelling Salesman Problem (TSP) for a given weighted graph (Naomi). In order to do this, I added or replaced edges to represent the lowest-weight pathways between vertices that were not directly connected, converting the practical TSP into a classical TSP. For example, vertex UB is traversed by the least-weight path from vertex UM to vertex UBr, so the path is UM → UB → UBr with a total weight of £55.00 (25.60 + 29.40). In this graph, the shortest routes between previously unconnected vertices consistently passed through vertex UB. To calculate these, I summed the weights of edges between each pair of vertices and vertex UB. This approach of adding edges to connect every vertex with all others resulted in the complete weighted graph in Figure 6 along with the corresponding adjacency table shown in table 5.

E - understanding comes through explanation. (some or good but accurate).

Figure 6: Shows complete weight graph

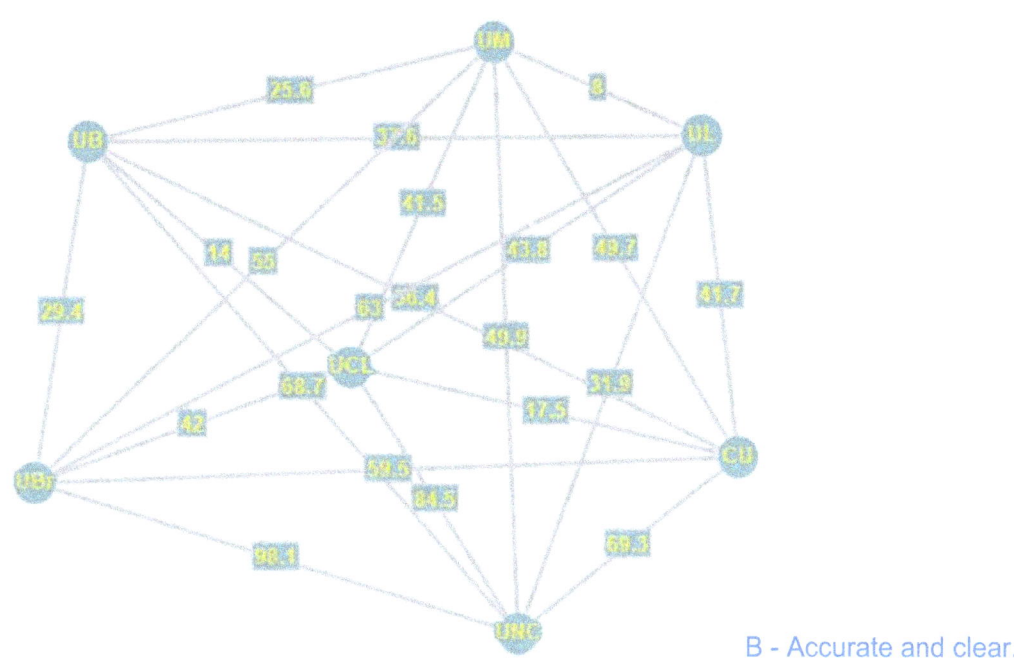

B - Accurate and clear.

Table 5: Complete weight adjacency table

Vertex	UCL	CU	UL	UM	UB	UNC	UBr
UCL		17.50	43.80	41.50	14.00	84.50	42.00
CU	17.50		41.70	49.7	36.40	69.30	59.50
UL	43.80	41.70		8.00	37.60	31.90	63.00
UM	41.50	49.7	8.00		25.60	49.90	55.00
UB	14.00	36.40	37.60	25.60		68.70	29.40
UNC	84.50	69.30	31.90	49.90	68.70		98.10
UBr	42.00	59.50	63.00	55.00	29.40	98.10	

With a complete weight/cost graph now established, I can find an approximate to answer the Traveling Salesman Problem (TSP). For a smaller graph with fewer vertices, it might be feasible to inspect all possible Hamiltonian cycles to identify the one with the minimum weight/cost. But, because of the number of vertices in my full-weight adjacency graph, this approach would be inefficient and impractical. Instead, I

D - Reflection of limitations. Reasoning good leading to educated decision.

will estimate the minimum weight/cost of the Hamiltonian cycle. To do this, I will need to determine both an upper and lower bound for the cycle's weight. The **Nearest Neighbour Algorithm** will be used to establish an upper bound.

The steps of this algorithm:

1. Select the starting vertex (for this graph, the starting vertex is UCL)
2. Follow the edge of least weight from the current vertex to an adjacent unvisited vertex (if there is more than one edge of least weight pick one at random)
3. Repeat step 2 until all vertices have been visited.
4. Add the final edge to return to the starting vertex

Based on the results from Table 5, I constructed another table (Table 6) using this algorithm, which shows the edges with the lowest weight/cost and their related weights. If a cycle is formed, the weight of that edge is ignored, and the next edge with the lowest weight is chosen in its place. Edges that would cause a cycle were disregarded for the rest of the table until every vertex had been visited once.

Table 6: Displays the Nearest Neighbor Algorithm

Edge of least the weight/cost	Weight/cost in British pounds (£)	Node/Vertex	Has a cycle formed?
N/A	N/A	UCL	N/A
UCL → UB	14.00	UB	No
UB → UM	25.60	UM	No
UM → UL	8.00	UL	No
UL → UNC	31.90	UNC	No
UNC → UBr	98.10	UBr	No
UBr → UCL	42.00	UCL	Yes

The Hamilton cycle deducted from the results in Table 6:

$$\text{UCL} \rightarrow \text{UB} \rightarrow \text{UM} \rightarrow \text{UL} \rightarrow \text{UNC} \rightarrow \text{UBr} \rightarrow \text{UCL}$$

The edge of the least weight/cost is shown above in a linear form.

Variable m is the minimum weight of the Hamiltonian cycle of the graph Figure 6, therefore the upper bound answer for the practical TSP using Figure 5.

$$m \leq 14.00 + 25.60 + 8.00 + 31.90 + 98.10 + 42.00$$

$$m \leq 219.60$$

With the upper bound m, now I need to find the lower bound for m. To do this the deleted vertex algorithm is used.

> E - section shows well-explained steps. Is precise with it's accurate calculations and easy to follow.

The steps of this algorithm:

1. Choose a vertex and delete it along with all edges that are connected to it
2. Deduce the **Minimum Spanning Tree** (MST) edge with the least weight for the rest of the graph.
3. Add the two shortest edges that were deleted from the original graph to the weight of the minimum spanning tree

First, I needed to select a vertex to be removed, and using a random generator, UBr was chosen.

Table 7: Displays Weight adjacency table with vertex UBr removed

Vertex	UCL	CU	UL	UM	UB	UNC
UCL		17.50	43.80	41.50	14.00	84.50
CU	17.50		41.70	49.7	36.40	69.30
UL	43.80	41.70		8.00	37.60	31.90
UM	41.50	49.7	8.00		25.60	49.90
UB	14.00	36.40	37.60	25.60		68.70
UNC	84.50	69.30	31.90	49.90	68.70	

The edges with the least weight in vertex UBr were UBr → UB and UBr → UCL, with weights of £29.40 and £42.00. These weights/costs will be added to the weight of the MST to calculate the lower bound.

With Vertex UBr removed, we can find the MST using **Prim's algorithm** This algorithm can be applied in two methods. The first method involves choosing the closest edges to each vertex by examining the graph. But, given that all vertices in the graph have an order of 6 (including Vertex UBr), making this approach is too complex. Therefore, I will use the second method by applying Prim's algorithm using a weighted adjacency table with the following steps:

> C - engages with alternate options before making an informed decision of what his next steps are.

1. Randomly choose a vertex and remove the row corresponding to the selected vertex.
2. Number the column of the selected vertex.
3. Circle the lowest undeleted entry in any of the numbered columns. If there are multiple choices, pick one at random. Else the new chosen vertex is selected by the row circled edge.
4. Repeat steps 1-3 until all rows are crossed out. The values in the red circle are the edges of the MST.

Prim's algorithm is shown in table 8 below. The deleted rows are represented by black horizontal lines.

Table 8: Displays the Prim's Algorithm

	1	6	4	3	2	5
Vertex	UCL	CU	UL	UM	UB	UNC
~~UCL~~		~~17.50~~	~~43.80~~	~~41.50~~	~~14.00~~	~~84.50~~
~~CU~~	(17.50)		~~41.70~~	~~49.7~~	~~36.40~~	~~69.30~~
~~UL~~	~~43.80~~	~~41.70~~		(8.00)	~~37.60~~	~~31.90~~
~~UM~~	~~41.50~~	~~49.7~~	~~8.00~~		(25.60)	~~49.90~~
~~UB~~	(14.00)	~~36.40~~	~~37.60~~	~~25.60~~		~~68.70~~
~~UNC~~	~~84.50~~	~~69.30~~	(31.90)	~~49.90~~	~~68.70~~	

Adding up the weight/cost of the edges circled in red, the MST weight can be calculated: 14.00 + 25.50 + 8.00 + 31.90 + 17.50 = £96.90   E - Accurate

With the MST calculated which is £96.90, next is to solve the deleted vertex algorithm. Do this, first, the length of the two shortest edges of the deleted vertex is added to £96.90. The shortest edges were UBr → UB and UBr → UCL, £29.40 + £42.00 = £71.40. The result of the deleted vertex algorithm represents the lower bound for the minimum weight needed to solve the Traveling Salesman Problem.

$$£96.90 + £71.40 = £168.30$$

$$m ≥ £168.30$$

$$∴ £168.30 ≤ m ≤ £219.60$$

The range of £168.30 ≤ m ≤ £219.60 for m gives me an estimate of the cost to visit all 7 universities by rail, starting and ending the train journey in London at vertex UCL.

E - Comes to a logical and precise answer to his second aim thanks to his understanding of the course.

Conclusion:

Evaluation of results:

In conclusion, it was determined that the University College London would be the optimal university with regards to location in proximity to the most nearby airports. This was expected because the university is located in the capital city of the UK. However, discovering that the University of Bristol has the second most number of airports, 5 in total, near its campus was surprising. This led me to question the effectiveness and accuracy of using Voronoi diagrams in this IA, illustrated in Figure 4. This questioning of the effectiveness of a Voronoi resulted in my use of the circles method to find the number of airports in a 91.75km radius. To answer my second aim, the solution to the practical Traveling Salesman Problem has given me a range of travel expenses from £168.30 to £219.60. This gives an idea of what the budget for my visit to each university on my shortlist will be. A limitation of this investigation is the accuracy of ticket price approximations, as train ticket prices can vary due to factors beyond my control. Therefore, while my TSP solution offers an accurate estimate for now, it may not stay as accurate over time in the future. In addition, the Nearest Neighbour Algorithm could result in a maximum error of

$1 + \frac{\log(n)}{2}$ times the overall cost of an optimal tour m, where n represents the number of vertices or universities. Consequently, the approximated cost given by the Nearest Neighbour Algorithm becomes less accurate as n increases.

> All good until the paragraph above. Should not be introducing new maths into the conclusion. That is for the main body.
>
> A and D: Conclusion in general offers good reflection D and relevance to the aims A completing the flo the exploration.

The maximum error is calculated in the following steps:

For my graph: n = 7

$$Maximum\ Error = \left(1 + \frac{\log(7)}{2}\right) \times m$$

$$Maximum\ Error = 1.423 \times m\ (3\ d.p.)$$

If I had narrowed down my university options to 5: n = 5

$$Maximum\ Error = \left(1 + \frac{\log(5)}{2}\right) \times m$$

$$Maximum\ Error = 1.349 \times m\ (3\ d.p.)$$

This suggests that my estimated cost for the TSP could have been more precise if I had narrowed down my selection of universities earlier, resulting in a graph with fewer vertices to identify for the Hamiltonian cycle.

Extension of IA: D - even at the end he is taking the time to further reflect and looking at how it could be improved.

Instead of using cost for the weight of the edges, travel time could be used or both. This would have given me the time it would take me to travel and the cost of visiting all of the shortlisted universities.

During this IA I have faced quite a few challenges when it came to selecting the right HL graph theory. Such as the problem with the classical TSP initially, I thought that it would work but after trying it I found out that it wouldn't, so I had to switch to the practical TSP. This made me change my graph theory algorithm 3 times before selecting the TSP graph theory this was because other graph theories were incompatible.

Bibliography:

Bellelli, Francesco. "The Fascinating World of Voronoi Diagrams." *Builtin*, 22 Feb. 2023, builtin.com/data-science/voronoi-diagram#:~:text=A%20Voronoi%20diagram%20is%20a,architecture%2C%20art%20and%20computer%20science.

Wolfram, Mathworld. "Voronoi Diagram." *Wolfram Mathworld*, https://mathworld.wolfram.com/VoronoiDiagram.html.

GeoGebra. "Graphing tool." *GeoGebra*, www.geogebra.org/m/qq92F3at.

Wolfram MathWorld. "Hamiltonian Cycle." *Wolfram MathWorld*, 12 Jan. 2024, mathworld.wolfram.com/HamiltonianCycle.html.

C, Naomi . "Graph Theory." *SaveMyExams*, 15 May 2024, www.savemyexams.com/a-level/further-maths_decision-maths-1/edexcel/17/revision-notes/algorithms-and-graph-theory/graphs/graph-theory/.

IA Sample 12 Examiner comments: Gini Coefficient

A very good IA where the student calculates the Gini Coefficient of Denmark and South Africa using modelling and integration.

Appropriate for: AA, SL, AAHL, AIHL   Possible for: AISL

Criterion	SL	HL
A	3	3
B	3	3
C	3	3
D	2	2
E	5	4
Total	16	15

A: The structure is logical and flows clearly from context to data collection, modelling, integration, and comparison. Tables and graphs are used effectively, with Lorenz curves and income shares well presented. The report is divided into manageable, well-labelled sections, though some transitions could have been done more smoothly. Some minor formatting issues. Flow of the IA could be improved with fewer headings.

B: Functions are written clearly, and the notation is mostly consistent. Variables are introduced and defined. Degree of accuracy needs to be mentioned. Some of the formatting could be improved.

C: The topic is interesting and relevant, and the student has clearly chosen it based on a meaningful social interest. There is good initiative in gathering data and using appropriate models for different countries. The student's comparison and interpretation of results show an awareness of broader implications. The student engages well with the mathematics and lets the results guide the exploration.

D: The student has four clear "Reflection" sections. This is not recommended to help with criterion A, but lots of reflection is evident. The student reflects briefly on the meaning of the Gini coefficient values and what they say about inequality in Denmark and South Africa. They acknowledge some limitations of the data. More reflection on model validity would help achieve 3/3.

E: Mathematics is commensurate with level of the course including construction of Lorenz curves using polynomial and exponential functions, and integration to find the area under the curve. Gini coefficient calculated accurately and interpreted correctly. The application is meaningful and shows good understanding of both the mathematics and its real-world context. Student demonstrates a good understanding of the maths. More thorough explanations of the modelling needed to achieve 6/6.

Introduction

The aim of this IA is to find the Gini coefficient of different countries with a focus on Denmark and South Africa by using integration and understanding why these two countries are different in income equality.

Economic development and growth is not necessarily associated with all people becoming richer. Income inequality is, and has been, a major issue to the world within macro income distribution. This issue of economic development in relation to distribution of wealth is especially a very interesting topic.

I am from Denmark which is known for having one of the most stable economies with relatively low growth whereas the United States of America (USA) has a huge economy experiencing relatively high growth. Therefore, I have had an interest to investigate Denmark, South Africa and other countries, one in particular being the US income distribution.

My interest began as a result of my Danish nationality. furthermore, I recently took a trip to South Africa and found the development of Cape Town in relation to other South African cities very fascinating. I Therefore, also wanted to consider South Africa in my IA. I also found the use of calculus in finding the equality of a country to also be interesting.

I considered comparing each country in the study to each other and analyzing the reason for the differences in income distributions. With that, I decided to model the gini index together with my own knowledge of integration. I find this combination interesting as Math's and Economics are currently my favorite subjects. I challenged myself in this report by using integration to calculate the area of the Lorenz Curve. So far, my math class has not yet studied integration, so everything was self-taught.

With the interest in hand I will be choosing to analyze four counties. These three countries will be Denmark, South Africa, United States due to US having one of the strong economies. My last country will also be England due to me wanting to study there in the future and having a better understanding of their economy.

Equity distribution of income is one of the subjects within in Macroeconomics and it is one of the five objectives for most governments. Them being stable economic growth, stable inflation rate, stable employment, stable balance of payment, and improve income distribution. However, it is almost impossible to accomplishes this goal for all governments. It is important to measure income equality in comparison to Economic Growth.

The basic's of the Gini Coefficient is very simple with if the Area of A from Diagram 1 is equals to 0 there is perfect income equality and if it equals to 1 then there is maximum possible income inequality. Therefore the higher it is from 0 to 1 the higher the possible income inequality.

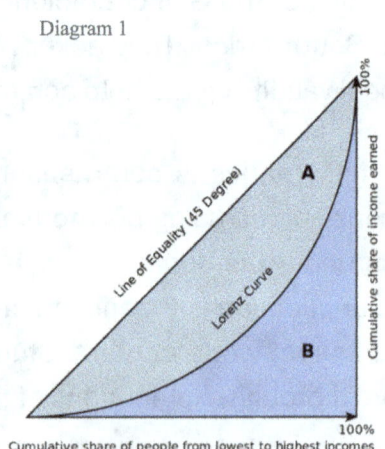

Diagram 1

Measuring income distribution is done with the Gini Coefficient and the Lorenz Curve. Diagram 1 shows the Lorenz curve as a visual representation of the share of income received by the percentage of the population. The Lorenz curve is constructed by gathering all data of income collected by households and then arranging them according to the level of income. Income is divided into 5 groups (quintiles) of equal sizes. This means that each group consists of 20% of the country's households. the lower the value is on the y axis, the lower the cumulative share of income earned and vice versa Looking at diagram 1 the x axis represents the cumulative increase in population from lowest to highest as a percentage. The y axis represents the household income in the country as a percentage seen in diagram 1 and shows the cumulative share of income earned. You calculate the cumulative share of income by adding the total share of income earned from each group. For example, graph 1 is taken from the income distribution of Denmark. The first group has a total share income of 9% and the next group has 13% of the income. This means that the first 20% of the population would have 9% and the first 40% would have 22% of the total income. After plotting all 5 points, then there is the 45-degree angle line which represents the line of perfect income distribution in diagram 1. Here, each group would earn 20% of the national income such that $f(x) = x$. The Lorenz curve is based on the quintile data. The Gini Index is based on the Lorenz curve and it shows income inequality within an economy. The area that is between the Lorenz curve and the line of perfect income equality is shown as A in diagram 1. This value is then divided by the total area of the right triangle.

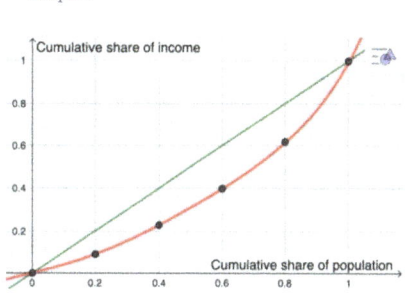

Graph 1

Hence, the formula used to calculate the area is

$$GI = \frac{A}{A+B}$$

Data Collection

In order to calculate the income distribution of the four countries selected I will be collecting the share of total income held by quintile data collected from the World Bank Database. The four countries being United Kingdom, USA, South Africa, and Denmark. The data I will be collecting is represented in quintiles of five equal parts, each representing in 20% of the total population as seen in Table 1 below.

Table 1: income held by each quintile for different country

Country	Quintile	Income share held (%)	Cumulative share of income (%)
Denmark(2017)	Bottom 20%	9.1	9.1
	Second 20%	13.7	22.8
	Third 20%	17.2	40
	Fourth 20%	21.9	61.9
	Top 20%	38.1	100
South Africa (2014)	Bottom 20%	2.4	2.4
	Second 20%	4.8	7.2
	Third 20%	8.2	15.4
	Fourth 20%	16.5	31.9
	Top 20%	68.2	100.1≈100
United States (2016)	Bottom 20%	5.1	5.1
	Second 20%	10.3	15.4
	Third 20%	15.3	30.7

	Fourth 20%	22.6	53.3
	Top 20%	46.8	100.1≈100
United Kingdom (2016)	Bottom 20%	7.1	7.1
	Second 20%	11.9	19
	Third 20%	16.4	35.4
	Fourth 20%	22.5	57.9
	Top 20%	42.1	100

Reflection 1

I have seen videos and tutorials on how to calculate the Gini index. With my research from the data collected from the world bank I noticed that South Africa and United State both had a total Cumulative share of income equaling 100.1. Here l needed to round it down to 100 as the cumulative shared income should always equal 100. I also realized that data from each country was from different points in time. With Denmark being the most recent one at 2017, whereas South Africa is from 2014, however United States and United Kingdom having the same year at 2016. Therefore, I wanted to collect data from 2014 as it matches the data from South Africa. However, when trying to do that the US didn't have data from 2014 and therefore made m chose the most recent data for each country.

After calculating the cumulative share of income held (by adding up the total income of each quintile), I will then construct the Lorenz curve for each country. This curve would then represent the distribution of income in each country.

I used GeoGebra to plot the distribution of income for each of the countries and then determined an equation for the line of best fit to mathematically formulate this distribution. I considered different models to see which best fit best represents the distribution of income.

For my calculations I have decided to use 3 significant figures for the final answer of each of my calculations. However, to make to answer more accurate, in my calculations, I will use as many significant figures as possibly.

Also, for the graphs I decided to use 0 to 1 instead of 0 to 100 for both of my x axis and y axis. This is becuaise GeoGebra didn't allow me to get the appropriate function for my graph when using 0-100. Therefore, the cumulative share income and population are going to be 0 to 1. What this means is that 20% is 0.2, 40% is 0.4, 60% is 0.6, 80% is 0.8, and 100% is 1.

Denmark

Image 1 Denmark Highlights

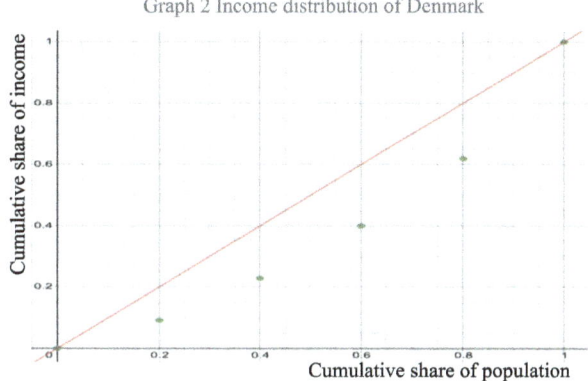

Before adding the line of best fit I will, as seen in graph 1, add all the points for Denmark's income distribution and form there I will be able to find the Lorenz curve. Based on the graph, the growth it appears to be exponential growth and I am therefor going to fit an exponential function for my Lorenz curve.

I first graphed the Gini coefficient for Denmark as they are said to have the lowest Gini coefficient. From looking at the graph 1, I decided to set set my function to $f(x) = a \times e^{bx} + c$ where I can then get the line of best fit to be represented as $0.1858875365e^{1.8334470071x} - 0.1684786419$ from the five points shown below

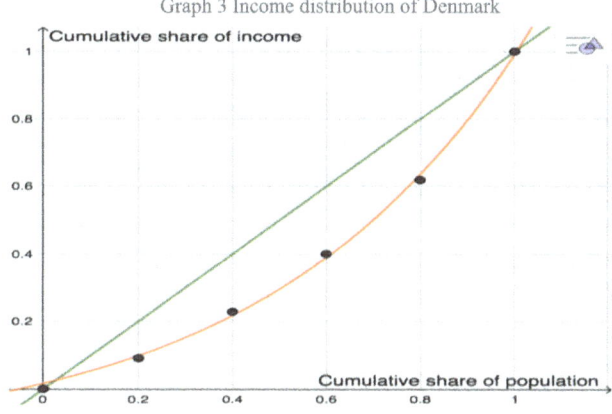

With the exponential function, it didn't intercept at (0,0) and (1,1). However, with the Gini Coefficient measuring from (0,0) and (1,1) I will assume these points as I am measuring from these points in regard to the population. The calculation of B is represented below:

$$f(x) = 0.1858875365e^{1.8334470071x} - 0.1684786419$$

Given the Lorenz curve of Denmark's I am gonna integrate the given function to find the area of B shown in diagram 1

We can then Apply the Sum rule to integrate the equation

$$\int_0^1 0.1858875365e^{1.8334470071x} - 0.1684786419 \, dx$$

Therefore;

$$\left[\frac{1}{1.8334470071}0.1858875365e^{1.8334470071x} - 0.1684786419x\right]_0^1$$

Where I then substitute to points of interest

$$= \left[0.101387e^{1.8334470071(1)} - 0.1684786419(1)\right]$$
$$- \left[0.101387e^{1.8334470071(0)} - 0.1684786419(0)\right]$$

$$0.634217 - 0.1684786419 - 0.101387$$

$$B = 0.364351$$

Knowing the area of B between point (0,0) and (1,1), because $A + B = 0.5$, the answer is shown below.

$$A + B = 1 \times 1 \times \frac{1}{2} = 0.5$$

After the area of B I am going to subtract it with 0.5 which would lead me to find what the area of A is. Calculation seen below.

$$A = 0.5 - B \approx 0.5 - 0.364351 = 0.135649 \approx 0.136$$

Given this, the Gini Index is given as:

$$G = \frac{A}{A+B} = \frac{0.135649}{0.5} = 0.271298 \approx 0.271$$

Reflection 2

After completing the calculation of the exponential function of Denmark from looking at graph 1 there is a clear understanding that the Lorenz curve does fit pretty well from the points collected. However, it doesn't intersect at the points of (0,0) and at (1,1). Therefore, to increase accuracy I will try to do the quintic function for describing Denmark's income distribution. The calculations will have 10 decimal points, however the below representation will be with 3 significant figures for each of my equations.

Good clear reflection

As before, I added the points from the data collected from Denmark to GeoGebra and set a polynomial function to five degrees make the Lorenz curve intersect (0,0) to (1,1). The equation provided by GeoGebra was given as:

$$f(x) = 2.08x^5 - 3.57x^4 + 1.97x^3 + 0.14x^2 + 0.37x$$

this can then be modeled as below:

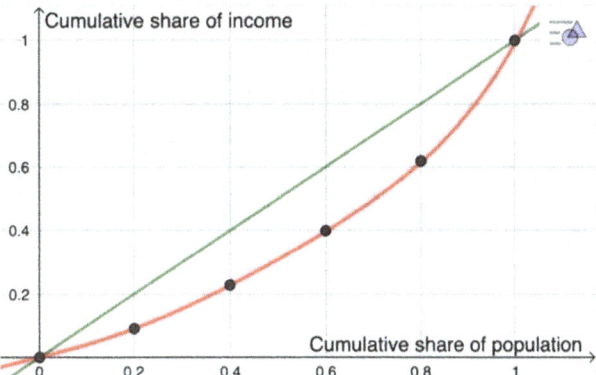

Graph 4 Income distribution of Denmark

B+ good clear graph

to find the area under the polynomial we can find the integration as followed:

$$\int_0^1 2.08x^5 - 3.57x^4 + 1.97x^3 + 0.14x^2 + 0.37x \, dx$$

Then we can apply the power rule for each function.

$$= \left[\frac{2.08x^6}{6} - \frac{3.57x^5}{5} + \frac{1.97x^4}{4} + \frac{0.14x^3}{3} + \frac{0.37x^2}{2}\right]_0^1$$

This can be simplified as

$$[0.3467x^6 - 0.714x^5 + 0.4925x^4 + 0.0467x^3 + 0.185x^2]_0^1$$

$$B = (0.3467(1)^6 - 0.714(1)^5 + 0.4925(1)^4 + 0.0467(1)^3 + 0.185(1)^2) - (0.3467(0)^6 - 0.714(0)^5 + 0.4925(0)^4 + 0.0467(0)^3 + 0.185(0)^2)$$

$$B = 0.3467 - 0.714 + 0.4925 + 0.0467 + 0.185 \approx 0.3569$$

By doing this we have then found the answer for Area B of Diagram 1. The line of equality is given as $f(x) = x$ and intersects at (0,0) and (1,1). Knowing this we can find A by just substituting it with 0.5.

$$A + B = 1 \times 1 \times \frac{1}{2} = 0.5$$

$$A = 0.5 - B \approx 0.5 - 0.3569 = 0.1431$$

Given this, I can then plug the answer for A being 0.1431 into the Gini Index formula.

$$G = \frac{A}{A+B} = \frac{0.1431}{0.5} = 0.286$$

Reflection 3
We can compare both the answers from the quintic function and the exponential function. The quintic function gave 0.2862 and the exponential function was 0.271. Both functions have a 0.01 difference which might be due to the exponential function having 10 decimal points and the quintic having only three significant figures. This difference is one of the flaws with my calculated Lorenz curve. If the exponential function had three significant figures, then the answer had a difference of 10 percent from Gini index derived by the world bank. Considering decimal points allowed for a much smaller

difference of about 1%. Given I have not use GeoGebra before, another problem I experienced was using it to find the best fit. I started with no idea on how to do it and I spent considerable time figuring it out.

The registered Gini Coefficient of Denmark in 2017 is 0.287 according the world bank. Which means that my value for the quintic function was off by 0.001 under the target index and my exponential function was off by 0.01 over the target index. This is likely because of economic theory of the Lorenz curve only exists between (0,0) and (1,1). My exponential function, however, didn't pass these points.

South Africa

Image 2 South Africa Highlights

Following the same steps as before, I again set the income distribution of South Africa into GeoGebra. I first set to function with be best fit to a polynomial with a five degree as such; $f(x) = ax^5 + bx^4 + cx^3 + dx^2 + gx$.the resulting function was determined as:

$$f(x) = 8.98x^5 - 16.95x^4 + 11.57x^3 - 2.97x^2 + 0.37x$$

As modelled below:

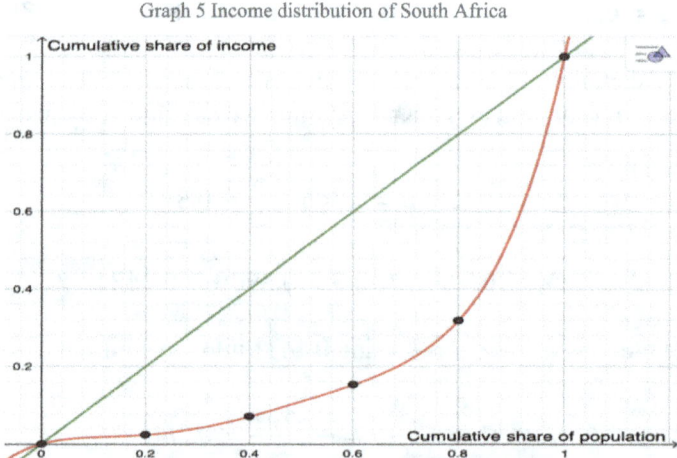

Before undergoing the calculation, based on just looking at the graph, there is a noticeable difference between South Africa and Denmark. There is a massive jump in cumulative share of income between 0.8 or 80% of the population, to1 or 100% of South African population.

The method of calculation is as showed under Denmark and the full process is shown in appendix 1.1. What can be seen from both of the quintic functions, South Africa has a much steeper curve than

Denmark, with the top 20% of South Africans holding the majority of total income. Seeing this easily shows that the income distribution is more unequitable in South Africa than in Denmark

The calculation from the quintic function can be seen in appendix 1.1 where we can determine a Gini Coefficient of 0.612. The answer clearly shows how income distribution in South Africa is unequitable compared to the Danish Gini Coefficient of 0.286.

With the regard to an exponential function, I set the function to $f(x) = a \times e^{bx} + c$ where then I get the line of best fit to be $0.0028211541 e^{5.8324784471x} + 0.0350185231$ from the five points as shown below

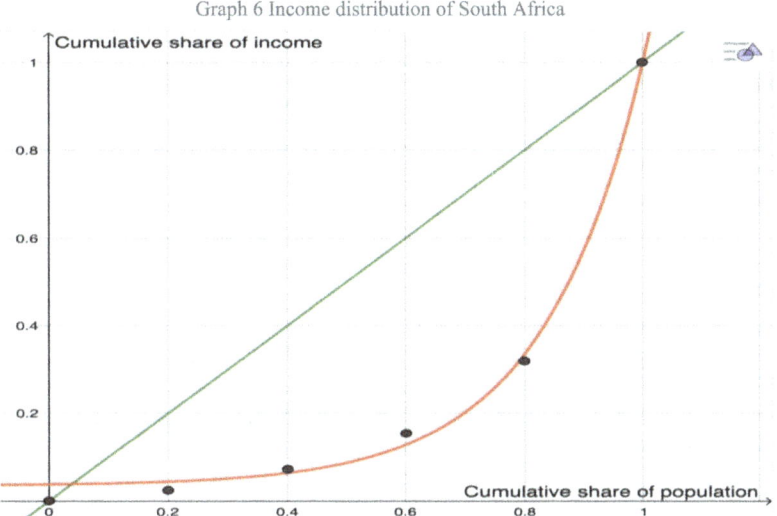

The calculation of the exponential function can be seen in appendix 1.2 where the Gini Coefficient is given as 0.598.

The Gini coefficient of South Africa is 0.63 in 2014 according to the World Bank. The difference of quintic from this value is 0.02 and with exponential function it is 0.03. the quintic is thus, closer to the determined Gini coefficient. The reason for this could be due to the collection of data regarding the cumulative share of income which added up to 100.1. I rounded this down to 100 as the total population of South Africa can't be 100.1% of population and should always sum to 100. This could be one of the problems as to why my data wasn't perfectly correct. The reason for why the exponential function is not the same as the quintic function could be due to, again, to the Lorenz curve needing to pass (0,0) and (1,1) because 0% of the population should earn 0% of the cumulative income and 100% of the population should earn 100% of cumulative income.

United States

Image 3 United States Highlights

The United States has one of the strongest economies in the world, however, does the income distribution correlate with the United States economy?

As before, I plotted the five points for the cumulative share of income and I set the function to be $f(x) = ax^5 + bx^4 + cx^3 + dx^2 + gx$ to get the equation for the best fit in GeoGebra. The equation is:

$$f(x) = 3.13x^5 - 5.6x^4 + 3.55x^3 - 0.29x^2 + 0.21x$$

and can be graphed below:

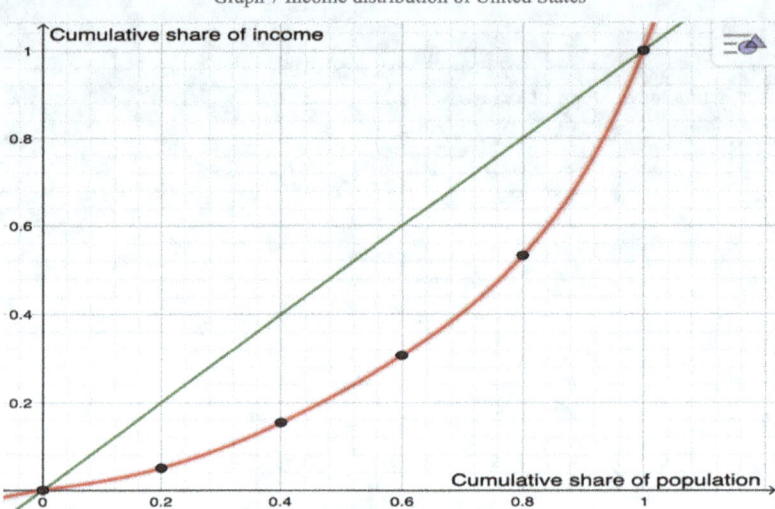

Graph 7 Income distribution of United States

By only looking at the graph it can already be estimated that United States Gini Coefficient will be smaller than that compared to South Africa. However, it should be larger than Denmark as it can be seen by the shape/slope of the Lorenz Curve. The Calculation can be seen in appendix 2.1

The answer seen in appendix 2.1 the quintic function getting a value of 0.405

With the exponential function I set my function to $f(x) = a \times e^{bx} + c$ where the line of best fit is shown as:

$$f(x) = 0.065784692e^{2.7691771517x} - 0.0533940185$$

from the five points shown below

Graph 8 Income distribution of United States

The calculation of the exponential function can be seen in appendix 2.2 where the Gini Coefficient is determined to be 0.397

According to the World Banks the Gini Coefficient of the United States is 0.415. Whereas the calculation from the quintic was 0.405 and with the exponential function determined a value of 0.397. With this, the difference of the quintic is 0.01 and exponential is 0.05, the closest one to the Gini Index is the quintic function. However, I believe this happened because I, again, had to round down from 100.1 to 100. Therefore, I believe that only a difference by 0.01 is more than accurate as the cumulative share income was 100.1.

United Kingdom

Image 3 United Kingdom Highlights

For the United Kingdom, I plotted the five points into GeoGebra and showed the five points of the United Kingdom given in a polynomial function. Which the given function was:

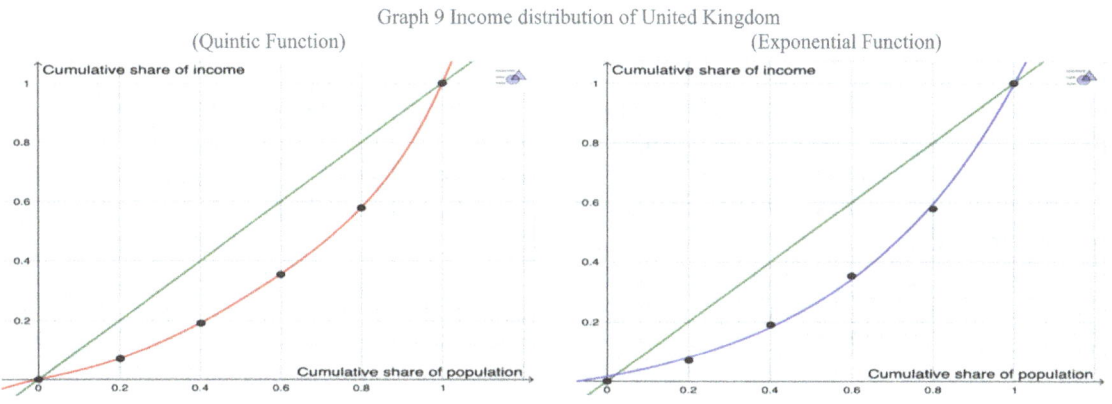
Graph 9 Income distribution of United Kingdom
(Quintic Function)   (Exponential Function)

Try to keep from going outside the margins

The graphs show the United Kingdom's income distribution is smaller than what I expected, however, it is probably still smaller than the United States. Both calculations are shown below:

The quintic function $f(x) = 2.6x^5 - 4.71x^4 + 2.99x^3 - 0.19x^2 + 0.31x$ The calculation for the quintic function can be seen in Appendix 3.1 from the quintic function is 0.34	From the exponential function I set my function to $f(x) = a \times e^{bx} + c$ and then I get the line of best fit to be: $f(x) = 0.1111043214 e^{2.2840902756x} - 0.0954608838$ The calculation for this function can be seen in Appendix 3.2 where the Gini coefficient was calculated to be 0.333

The World Bank Gini Index of the United Kingdom is 0.34. The calculation from the quintic was 0.34 and with the exponential function was 0.333. With that, we can see that the quintic was most accurate however the exponential function was under the target by 0.01. The calculation that was gotten from the quintic function was the same as the Gini index according to the world bank.

Reflection 4

The calculation was relatively easy to calculate as all the quintic functions intercepted at (0,0) and (1,1) however the exponential function did not, but I continued to integrate from 0 and 1 as I am measuring from the population of 0 to 100 (100 being 1). From there on, by looking at the compartment of the calculation of the quintic function, the outcome was either the same or very close to the Gini Index determined from the world bank. I was quite surprised when the US had a higher Gini Index than UK considering that US have one of the strongest economies. Originally, I thought both countries would have the same Gini Index.

After the calculation of all the four countries we can look at which function best represented each country in term of the exponential or the quintic function. This can be shown in table 1 with each of the answers we determined.

Table 1 Calculation compartment between each equation.

Countries	Exponential Function Answer	Quintic Function Answer	Gini Index
Denmark	0.271	0.286	0.287
South Africa	0.598	0.612	0.631
United States	0.397	0.405	0.415
United Kingdom	0.333	0.340	0.348

Looking at table 1, I had 2 different results from each country however I got the most accurate results from the quintic function. However, I still wanted to consider the exponential as all countries showed exponential growth. the quintic function was however more accurate. The Gini Index of each country that was collect to compare was from the world banks website.

Conclusion

In conclusion, the Gini Coefficient and the Lorenz curve are useful to calculate income distribution between the four countries. I was able to calculate the Gini index of each country by using calculus to find the answer. This is also interesting as I learnt how to properly calculate the Gini Index. Doing so is not taught in both economics and mathematics at the IB so it was interesting to calculate the Gini index using integration.

The values we determined were relatively accurate compared to the fact that they were only off by 0-5% compared to the World Bank results. We can also see that the difference between each country is very clear. Just by looking at the graph of South Africas income distribution compared to Denmarks there is a noticeable difference. this became even more clear with the use of integration, as the Danish Gini index was 0.286 and South Africa's was 0.612, which is a pretty big difference. What was found was Denmark's has the best income distribution compared to the three countries. As Denmark there is a lot of income distribution through the taxing system therefore a less income differences which would lead to a better Gini coefficient. Regards to South Africa is completely different

One of the challenges faced throughout the entire investigation was the recency of the data collected. For example, data collected on the UK and the US was from less than four years ago, data on Denmark was from 3 years, and South Africa's most recent data was from 2014. This Data could be have changed easily changed overtime as the economy changes. Therefore, this could be a problem when comparing Denmark and South Africa as the results are from 2014 and 2017. Another issue that I faced with the data was that I had to round up/down on many occasions where it may have been more accurate to consider more decimal places. To achieve this, however, I would need to use infinite numbers thereby making the calculations much more complicated.

To further extend, one could investigative the factors that effect the Gini Index. Possibly by consider the statistics correlation between the Gini index and other like taxes, expenditure education. Expenditure on health minimum wage that change income equality. we would probably expect the progressive tax system on Denmark and it's expenditure on social welfare.

Appendix
1.1: Calculating the Gini coefficient of South Africa using quintic functions

$$f(x) = 8.98x^5 - 16.95x^4 + 11.57x^3 - 2.97x^2 + 0.37x$$

$$\int_0^1 8.98x^5 - 16.95^{x^4} + 11.57x^3 - 2.97x^2 + 0.37x \, dx$$

> Good idea to put all these calculations in the appendices as it is repetitive

$$= \left[\frac{8.98x^6}{6} - \frac{16.95x^5}{5} + \frac{11.57x^4}{4} - \frac{2.97x^3}{3} + \frac{0.37x^2}{2}\right]_0^1$$

$$[1.4967x^6 - 3.39x^5 + 2.893x^4 - 0.99x^3 + 0.185x^2]_0^1$$

$$B = (1.4967(1)^6 - 3.39(1)^5 + 2.893(1)^4 - 0.99(1)^3 + 0.185(1)^2) - (1.4967(0)^6 - 3.39(0)^5 + 2.893(0)^4 - 0.99(0)^3 + 0.185(0)^2)$$

$$B = 1.4967 - 3.39 + 2.893 - 0.99 + 0.185 \approx 0.1947$$

$$A + B = 1 * 1 * \frac{1}{2} = 0.5$$

$$A = 0.5 - B \approx 0.5 - 0.1947 = 0.306$$

$$G = \frac{A}{A+B} \quad \frac{0.306}{5} = 0.612$$

1.2: Calculating the Gini coefficient of South Africa using exponential function.

$$f(x) = 0.0028211541e^{5.8324784471x} + 0.0350185231$$

$$\int_0^1 0.0028211541 e^{5.8324784471x} + 0.0350185231 \, dx$$

$$= \left[\frac{1}{5.8324784471} 0.0028211541 e^{5.8324784471x} + 0.0350185231x \right]_0^1$$

$$= \left[0.000484 e^{5.8324784471(1)} + 0.0350185231(1) \right]$$
$$- \left[0.000484 e^{5.8324784471(0)} + 0.0350185231(0) \right]$$

$$= 0.165142 + 0.0350185231 - 0.000484$$

$$B = 0.200645$$

$$A + B = 1 \times 1 \times \frac{1}{2} = 0.5$$
$$A = 0.5 - B \approx 0.5 - 0.200645 = 0.299355 \approx 0.299$$

$$G = \frac{A}{A+B} = \frac{0.299}{0.5} = 0.598$$

2.1: Calculating the Gini coefficient of United States using quintic functions

$$f(x) = 3.13x^5 - 5.6x^4 + 3.55x^3 - 0.29x^2 + 0.21x$$

$$\int_0^1 3.13x^5 - 5.6x^4 + 3.55x^3 - 0.29x^2 + 0.21x \, dx$$

$$= \left[\frac{3.13x^6}{6} - \frac{5.6x^5}{5} + \frac{3.55x^4}{4} - \frac{0.29x^3}{3} + \frac{0.21x^2}{2} \right]_0^1$$

$$[0.521667x^6 - 1.12x^5 + 0.8875x^4 - 0.096667x^3 + 0.105x^2]_0^1$$

$$= (0.521667(1)^6 - 1.12(1)^5 + 0.8875(1)^4 - 0.096667(1)^3 + 0.105(1)^2) - (0.521667(0)^6$$
$$- 1.12(0)^5 + 0.8875(0)^4 - 0.096667(0)^3 + 0.105(0)^2)$$

$$B = 0.521667 - 1.12 + 0.8875 - 0.096667 + 0.105 = 0.2975$$

$$A + B = 1 * 1 * \frac{1}{2} = 0.5$$

$$A = 0.5 - B \approx 0.5 - 0.2975 = 0.2025$$

$$G = \frac{A}{A+B} = \frac{0.2025}{0.5} = 0.405$$

2.2: Calculating the Gini coefficient of United States using exponential function

$$f(x) = 0.065784692 e^{2.7691771517x} - 0.0533940185$$

$$\int_0^1 0.065784692 e^{2.7691771517x} - 0.0533940185 \, dx$$

$$\left[\frac{1}{2.7691771517} 0.065784692 e^{2.7691771517x} - 0.0533940185x\right]_0^1$$

$$[0.023756 e^{2.7691771517(1)} - 0.0533940185(1)]$$
$$- [0.023756 e^{2.7691771517(0)} - 0.0533940185(0)]$$

$$0.378802 - 0.023756 - 0.0533940185 = 0.301652$$

$$A + B = 1 * 1 * \frac{1}{2} = 0.5$$

3.1: Calculating the Gini coefficient of United Kingdom using quantic functions

$$f(x) = 2.6x^5 - 4.71x^4 + 2.99x^3 - 0.19x^2 + 0.31x$$

$$\left[\frac{2.6x^6}{6} - \frac{4.71x^5}{5} + \frac{2.99x^4}{4} - \frac{0.19x^3}{3} + \frac{0.31x^2}{2}\right]_0^1$$

$$[0.433x^6 - 0.942x^5 + 0.747x^4 - 0.063x^3 + 0.155x^2]_0^1$$

$$(0.433(1)^6 - 0.942(1)^5 + 0.747(1)^4 - 0.063(1)^3 + 0.155(1)^2) - (0.433(0)^6$$
$$- 0.942(0)^5 + 0.747(0)^4 - 0.063(0)^3 + 0.155(0)^2)$$

$$B = 0.433 - 0.942 + 0.747 - 0.063 + 0.155 = 0.33$$

$$A + B = 1 * 1 * \frac{1}{2} = 0.5$$

$$A = 0.5 - B \approx 0.5 - 0.33 = 0.17$$

$$G = \frac{A}{A+B} = \frac{0.2025}{0.5} = 0.34$$

3.2: Calculating the Gini coefficient of United Kingdom using exponential functions

$$f(x) = 0.1111043214 e^{2.2840902756x} - 0.0954608838$$

$$\int_0^1 0.1111043214 e^{2.2840902756x} - 0.0954608838 \, dx$$

$$\left[\frac{1}{2.2840902756} 0.1111043214 e^{2.2840902756x} - 0.0954608838x\right]_0^1$$

$$[0.048643 e^{2.2840902756(1)} - 0.0954608838(1)]$$
$$- [0.048643 e^{2.2840902756(0)} - 0.0954608838(0)]$$

$$B = 0.477516 - 0.048643 - 0.0954608838 = 0.333412$$

$$A + B = 1 * 1 * \frac{1}{2} = 0.5$$

$$A = 0.5 - B \approx 0.5 - 0.333412 = 0.166588$$

$$G = \frac{A}{A+B} = \frac{0.166588}{0.5} = 0.333176 \approx 0.333$$

Citation
- Yglesias, Matthew. "Everything You Need to Know about Income Inequality." *Vox*, Vox, 7 May 2014, www.vox.com/2014/5/7/18076944/income-inequality.

- Chappelow, Jim. "Gini Index Definition." *Investopedia*, Investopedia, 29 Jan. 2020, www.investopedia.com/terms/g/gini-index.asp.

- "Income Share Held by Lowest 20% - Denmark, South Africa, United Kingdom, United States." *Data*, data.worldbank.org/indicator/SI.DST.FRST.20?locations=DK-ZA-GB-US&view=chart.

- "Income Share Held by Second 20% - Denmark, South Africa, United Kingdom, United States." *Data*, data.worldbank.org/indicator/SI.DST.02ND.20?locations=DK-ZA-GB-US.

- "Income Share Held by Third 20% - Denmark, South Africa, United Kingdom, United States." *Data*, data.worldbank.org/indicator/SI.DST.03RD.20?locations=DK-ZA-GB-US.

- "Income Share Held by Fourth 20% - Denmark, South Africa, United Kingdom, United States." *Data*, data.worldbank.org/indicator/SI.DST.04TH.20?locations=DK-ZA-GB-US.

- "Income Share Held by Highest 20% - Denmark, South Africa, United Kingdom, United States." *Data*, data.worldbank.org/indicator/SI.DST.05TH.20?locations=DK-ZA-GB-US.

- "GINI Index (World Bank Estimate) - United States, United Kingdom, South Africa, Denmark." *Data*, data.worldbank.org/indicator/SI.POV.GINI?locations=US-GB-ZA-DK.

IA Sample 13 Examiner Comments: Diwali Lights

Appropriate for: AAHL, AIHL

Criterion	SL	HL
A	4	4
B	4	4
C	3	3
D	3	3
E	6	5
Total	20	19

A: The exploration is very well written. It is coherent and well organised. Considering the mathematics is beyond the syllabus, it is understandable for a peer. Comments on paper but multiple examples of excellent organisation. The IA has a nice flow and always maintains focus on aim.

B: The mathematical communication is appropriate, relevant, and consistent throughout. Multiple examples noted on paper. All terminology is correct. All graphs are appropriate with axes labelled. Variables are defined. Rounding is explained and consistent. I struggled to find any flaw.

C: The student has put together an outstanding, original and creative piece of work. Their personal engagement easily comes through when reading the paper. They clearly have a strong interest in their work. The student's personal engagement drives the exploration forward on numerous occasions which are noted on the paper.

D: There are many examples of significant reflection and some examples of critical reflection. The reflection steers the exploration, and you can clearly see the student thinking about what they have done and what they are going to do next. There are many examples marked on the paper.

E: The mathematics is commensurate with the course and beyond. The parts that are beyond the syllabus (and those within) are well understood and very well explained. The student clearly has a thorough understanding of the mathematics. Formulae are derived where appropriate and applied correctly. Rigour and sophistication are demonstrated on multiple occasions, some noted on paper.

Introduction

Diwali is the Hindu festival of lights, representing the victory of light over darkness in stories I used to read as a child. One of my favourite traditions in Diwali is decorating my house with my family using festive string lights. The aim of this exploration is to find the length of string light I need to form a desired pattern, along with finding the area under this pattern where I can place flowers, using calculus. I will do this by designing the lights based on specifications, where they will be held on the wall at fixed points. To achieve the aim, I will model my design for the string light pattern using polynomials, specifically 5 quadratic equations. I will then model the pattern using catenaries, which are curves forming the shape of a flexible chain when attached at two fixed points (Carlson, 2017). Catenary curves can be formed by hyperbolic cosine functions, such as $y = \cosh x$ (Carlson, 2017).

Once both patterns have been modelled, I will use the arc length formula to find how much string light is needed for each pattern. After this, the area under the string lights will be found, as I want to decorate the rest of my wall with flowers, and I will need to find out how many to buy. I am interested to see how I can use calculus with hyperbolic functions, so I will attempt to integrate analytically throughout the exploration to calculate these lengths and areas.

Specifications

Table 1 shows the required specifications for the pattern I want to design, where all values are in meters because the string lights sold are measured in meters.

Width of Wall (m)	1.7	
Height of Wall (m)	1.0	
	5 Alternating String Lights Curves	
	3 Small Curves	2 Large Curves
Horizontal Distance between fixed points (m)	0.3	0.4
Vertical Height (m)	0.225	0.4
Area of one flower (m²)	0.0025	

Table 1: Specifications for Diwali Decorations

Using the specifications, I have created a diagram of the area on my wall showing the string lights and flowers in Figure 1.

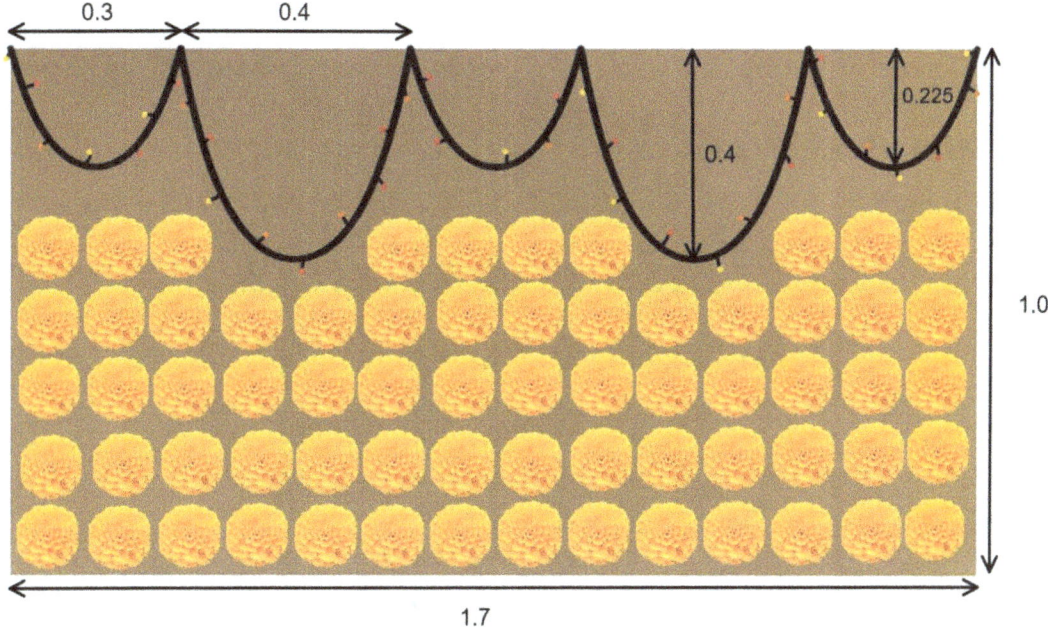

Figure 1: Diagram of Diwali Decorations (all values in metres). Flower image from: (Rawpixel, n.d.)

I will design models for the string lights based on these measurements, where I want to have 5 alternating curves, shown by the pattern above. I have specifically chosen these specifications since my family and I think they will look best on the 1.7m x 1.0m area on my living room wall, while leaving space for the flowers I am going to add underneath the lights. I measured the area of one flower by finding the length of its longest side (5cm) and squaring this, as each flower will assume the area of a square, since they are arranged in a grid. This will also allow me to space out the string lights creatively to form an appropriate pattern.

Designing the pattern with polynomials

Using these specifications and the nature of how strings hang on two fixed points, my initial thought for the string lights was that it clearly formed the shape of a quadratic, which I have studied deeply in school. Quadratic equations can be presented in the form:

$$f(x) = r(x - p)(x - q)$$

This means x-intercepts are at $(p, 0)$ and $(q, 0)$, where the constants $p, q, r \in \mathbb{R}$. We can then find the minimum, which will be used to find r when given values of p and q. The x co-ordinate of the minimum is the midpoint between both roots, so $x = \frac{p+q}{2}$. Substituting this value for x in the original function $f(x)$ gives the y co-ordinate of the minimum. Therefore, the co-ordinates of the minimum in the form (x, y) are $\left(\frac{p+q}{2}, r\left(\frac{(p-q)^2}{4}\right)\right)$, seen graphically in Figure 2 below.

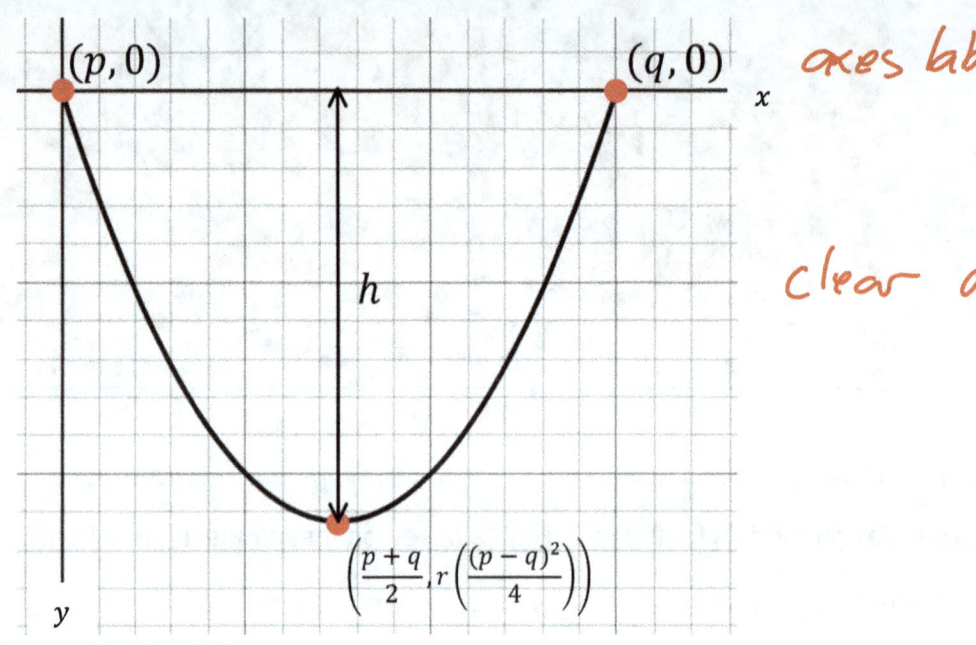

Figure 2: Finding r using properties of quadratics

My wall in the living room is 1.7m long, meaning the domain of the modelled function will be $\{0 \leq x \leq 1.7\}$. Splitting this into five sections, the 3 smaller curves will each have widths 0.3m and the 2 larger curves will have widths 0.4m at the roots, as shown by the specifications in Table 1.

I will make the wall start from the origin, so for the first curve, $p = 0$ and $q = 0.3$. We know the vertical heights of the curves from the specifications, where the height, $h = 0.225$ for small curves, $h \in \mathbb{R}^+$. Using this and the y co-ordinate of the minimum, we can find r through forming the equation: $h = r\left(\frac{(p-q)^2}{4}\right)$, where h is the vertical height in the specifications. Substituting in the values we know for p, q and h in the equation gives $r = 10$ for small curves. For the second, larger curve we can follow the same process with values: $p = 0.3$, $q = 0.7$, $h = 0.4$, also surprisingly giving the value $r = 10$.

Repeating this process and finding r for the remaining three curves, we can form the piecewise function:

$$f(x) = \begin{cases} 10(x)(x-0.3), & 0 \leq x \leq 0.3 \\ 10(x-0.3)(x-0.7), & 0.3 \leq x \leq 0.7 \\ 10(x-0.7)(x-1), & 0.7 \leq x \leq 1 \\ 10(x-1)(x-1.4), & 1 \leq x \leq 1.4 \\ 10(x-1.4)(x-1.7), & 1.4 \leq x \leq 1.7 \end{cases}$$

This function can then be presented on Desmos, shown in the graph below:

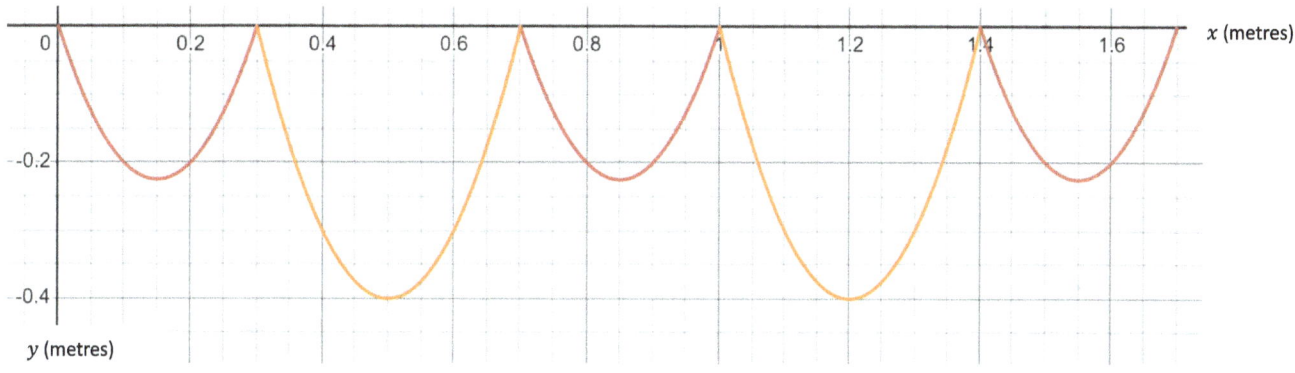

Figure 3: Diwali lights (modelled using quadratics)

Figure 3 shows the Diwali lights which will be placed on the area on my living room wall where one corner is (0,0) and another corner is (1.7,0). The roots, p and q, for each function are different based on the curve's position. The y-values for the minimum points are the same for all three small curves at $y = -0.225$, and for both large curves, $y = -0.4$, so we can see the heights are just as required by the specifications.

These y-values are negative since I have modelled the attachment points to be on the x-axis, to clearly show the curves are attached on a fixed horizontal line. It is also simpler to start the pattern at (0,0), the top-left of the wall. This will also be done for other patterns designed.

By looking at the model visually, I realised that it looks similar to a hanging chain, but I found that the slopes of the curves near the minima looked a little steeper in Figure 3 than in real life. This could negatively affect the accuracy of the exploration when using the model to calculate the final lengths and areas. Interested by how I could model the Diwali lights in another way, I started researching about the mathematics and physics behind hanging cables.

Designing the pattern with catenary curves

My perspective for this exploration changed when I learnt about catenary curves during my research. I realised that this may be a more accurate representation of the Diwali lights instead of quadratics, as each individual curve in the pattern is fixed between two points, hanging freely due to its own weight – the definition of a catenary curve (Carlson, 2017). Hyperbolic functions are commonly used to represent these curves, which have properties and identities like trigonometric (or circular) functions, although are modelled using hyperbolae (Carlson, 2017). A common hyperbolic function is the hyperbolic cosine function, $y = \cosh x$, shown in Figure 4, which can be used to model these catenary curves (Whitman College, n.d.). I found these fascinating since they are derived using exponentials instead of trigonometric functions (given their name), where the function $\cosh x = \frac{e^x + e^{-x}}{2}$ (Whitman College, n.d.).

The function is different to common trigonometric function $y = \cos x$, but I was interested in why they share similar names. I found that this was due to Euler's formula: $e^{ix} = \cos x + i \sin x$, where i is the imaginary unit (Newcastle University, n.d.). The formula represents the relationship between exponentials and standard trigonometric functions. To investigate further, I derived $\cos x = \frac{e^{ix} + e^{-ix}}{2}$ (see Appendix A). When the imaginary unit is removed, the two functions are similar in definition, hence why they have similar names.

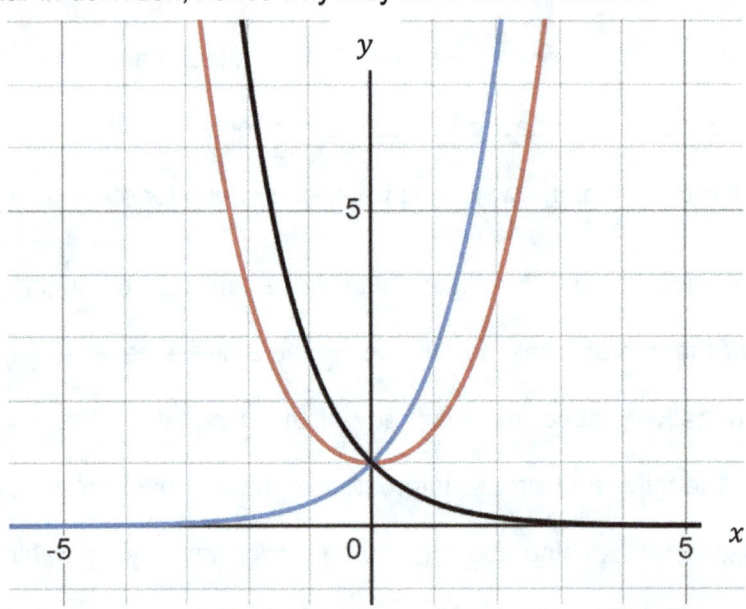

Figure 4: Hyperbolic cosine function, y = cosh x (red) made from y = e^x (blue) and y = e^-x (black)

Using Figure 4 and $\cosh x = \frac{e^x + e^{-x}}{2}$, we can see the hyperbolic cosine function is the mean of $y = e^x$ and $y = e^{-x}$, where all equations have a y-intercept with co-ordinates $(0, 1)$. We can see $y = \cosh x$ is symmetrical across the y-axis, therefore it is an even function, where $\cosh(x) = \cosh(-x)$. I wanted to see how these hyperbolic functions compared to quadratics as well, so in Figure 5 we can see $y = \cosh x$ being compared to quadratic $y = x^2 + 1$.

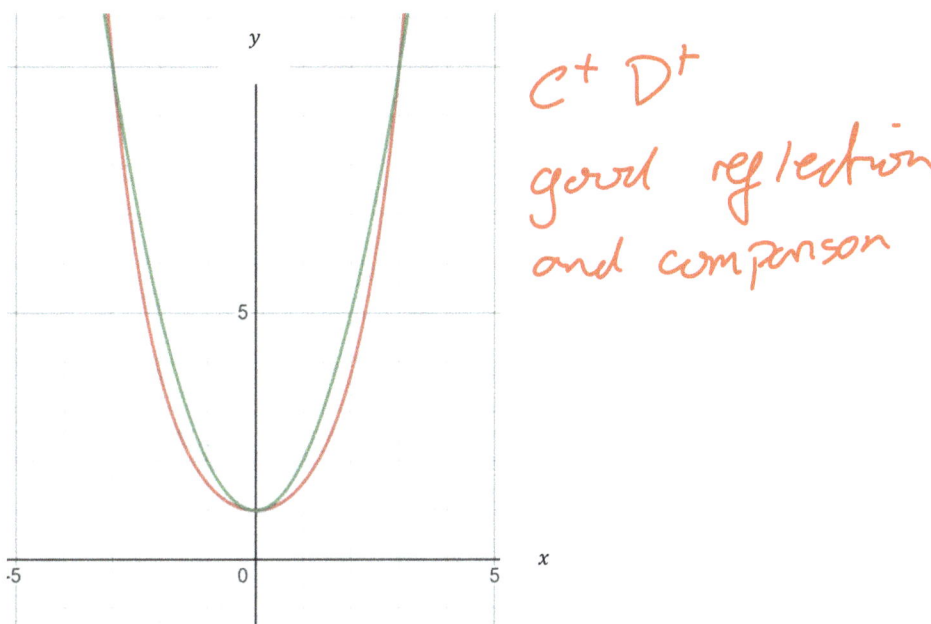

Figure 5: Comparison between hyperbolic cosine function and quadratic, $y = \cosh x$ (red) and $y = x^2 + 1$ (green)

I initially thought that hyperbolic functions would share no similarities with quadratics, however Figure 5 shows the shapes for these functions are quite similar, where both are concave upward. I found that the gradient of $y = \cosh x$ is less steep than the quadratic close to the minimum, meaning there is a slower rate of change at these points for $y = \cosh x$. I also learnt that a hanging cable, like a curve made from string lights, takes a shape more comparable to $y = \cosh x$ than the quadratic, because of the nature of the forces acting on the string light (Carlson, 2017). The forces of tension and weight must balance since the object is stationary, a concept in Newton's First Law. To explore this further, I read a report on the shape of hanging cables, where I learnt that these Diwali lights will assume a shape which minimises their potential energy (Georgia Institute of Technology, 2018). This is given by the shape of $y = \cosh x$ and a scaling factor, which is experimented with in Figure 6 (Georgia Institute of Technology, 2018).

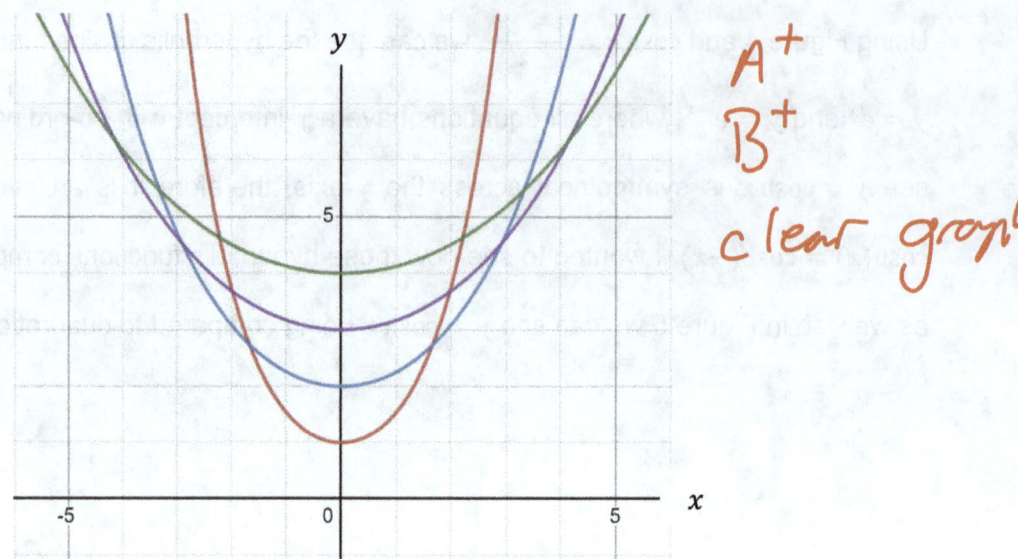

Figure 6: The Scaling Factor a, $y = \cosh x$ (red), $y = 2\cosh\frac{x}{2}$ (blue), $y = 3\cosh\frac{x}{3}$ (purple), $y = 4\cosh\frac{x}{4}$ (green)

An equation which models a catenary curve in cartesian co-ordinates is $y = a\cosh\left(\frac{x}{a}\right)$, where $a \in \mathbb{R}$ (Carlson, 2017). The constant a is the lowest y-value of the curve before translation, changing both horizontal and vertical scale factors. Increasing a reduces rate of change close to the minimum, where the minimum is $(0, a)$. Since this curve will be translated, we can form:

$$g(x) = a\cosh\left(\frac{x-b}{a}\right) + c$$

The constant b shows horizontal translation and c shows vertical translation, where $b, c \in \mathbb{R}$, shown graphically below. To use this equation to model the Diwali lights, we need to find these values. I want these catenary curves to have the same specifications as the quadratics.

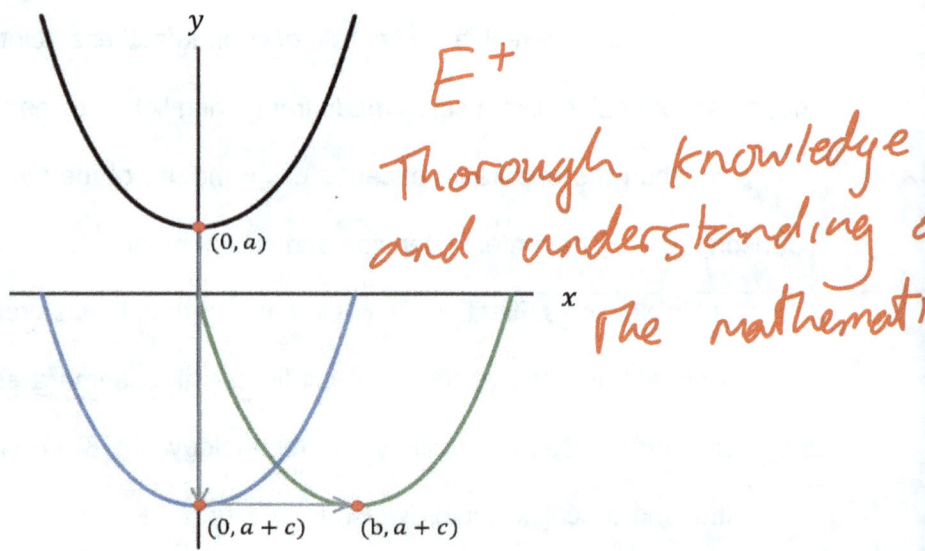

Figure 7: Horizontal and Vertical translation of curves

For all curves the minimum point will be $(b, a + c)$, since c will translate the curve vertically and b will translate the curve horizontally, away from the origin, seen specifically in Figure 7. For the same sized curves, a and c remain the same since the y co-ordinate of the minimum remains the same, although b changes as they are horizontally translated. For the first small curve (on the far left), $b = 0.15$, since it needs to be in the centre of x-intercepts $(0,0)$ and $(0.3,0)$ horizontally, and for the first large curve $b = 0.5$, since it is horizontally the midpoint of $(0.3,0)$ and $(0.7,0)$ as we know the catenary has an axis of symmetry, similar to quadratics. Continuing this, for the two other small curves $b = 0.85$ and $b = 1.55$, and for the other large curve $b = 1.2$, which will be used later. All values for the horizontal translation b will be applied after finding a and c.

Coherent clear explanations A+

Using the value a for small curves (a_s) and vertical translation c for small curves (c_s), we can show $a_s + c_s = -0.225$, where -0.225 is the y co-ordinate of the minima based on the height 0.225m required from specifications. I realised that the distance between roots needs to be 0.3m and there is perfect vertical symmetry in the line of the minimum point, so the roots will be $x = 0.15$ and $x = -0.15$, horizontally equidistant from the y-axis after translated downward (but before horizontal translation b). This is shown graphically in Figure 8:

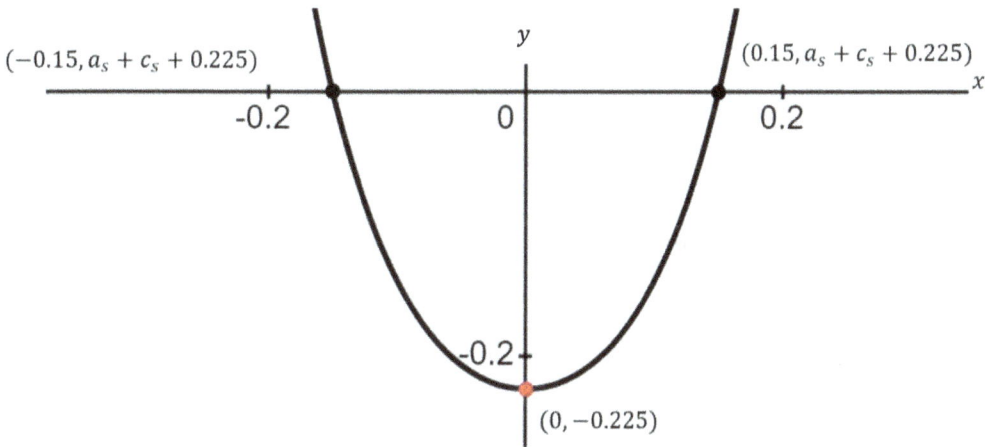

Figure 8: Finding co-ordinates of the roots (before horizontal translation applied)

Using the roots and $a_s + c_s = -0.225$, we can show that the co-ordinates of roots are $(0.15, a_s + c_s + 0.225)$ and $(-0.15, a_s + c_s + 0.225)$, as shown on Figure 8. To find a_s, we can use symmetrical properties of $y = \cosh x$, as it is an even function (see Figure 4).

Using these co-ordinates and $g(x)$, we can show (horizontal translation b is not applied yet):

$$y = a_s \cosh\left(\frac{x}{a_s}\right) + c_s$$

Rearranging for x:

$$x = a_s \cosh^{-1}\left(\frac{y - c_s}{a_s}\right)$$

Substituting the co-ordinates $(0.15, a_s + c_s + 0.225)$ and cancelling c_s gives:

$$0.15 = a_s \cosh^{-1}\left(\frac{a_s + 0.225}{a_s}\right)$$

Since the width between roots is 0.3m, we can double both sides:

$$0.3 = 2a_s \cosh^{-1}\left(\frac{a_s + 0.225}{a_s}\right)$$

We can then obtain the value $a_s = 0.071352\ldots \approx 0.071$ (3 d.p.). As $a_s + c_s = -0.225$, we find $c_s = -0.296352\ldots \approx -0.296$ (3 d.p.), which is negative, although it makes sense since the curve is translated downwards. These values have been rounded to 3 decimal places because I want measurements to be to the nearest millimetre when measuring how much is needed using a meter stick. I also want to be consistent, so I will keep rounding and displaying values to 3 decimal places throughout the exploration where required, but I will calculate the lengths and areas using the exact values for accuracy.

For the first small curve, using the values for a_s and c_s, along with $b = 0.15$ and $g(x)$, we can form (all values exact or to 3.d.p):

$$y = a_s \cosh\left(\frac{x - b}{a_s}\right) + c_s$$

$$y = 0.071 \cosh\left(\frac{x - 0.15}{0.071}\right) - 0.296 \quad \{0 \leq x \leq 0.3\}$$

Following this same process but using the value a for large curves (a_l) and vertical translation c for large curves (c_l), we have $a_l + c_l = -0.4$, which means the co-ordinates of the roots are $(0.2, a_l + c_l + 0.4)$ and $(-0.2, a_l + c_l + 0.4)$, as the function is symmetrical in the y-axis. We can then form the equation (before horizontal translation):

$$y = a_l \cosh\left(\frac{x}{a_l}\right) + c_l$$

Rearranging for x, substituting the co-ordinate $(0.2, a_l + c_l + 0.4)$ and cancelling c_l gives us:

$$0.2 = a_l \cosh^{-1}\left(\frac{a_l + 0.4}{a_l}\right)$$

Doubling both sides:

$$0.4 = 2a_l \cosh^{-1}\left(\frac{a_l + 0.4}{a_l}\right)$$

Then, we find $a_l = 0.081087\ldots \approx 0.081$ (3 d.p.). Additionally, we can use $a_l + c_l = -0.4$ to obtain $c_l = -0.481087\ldots \approx -0.481$ (3 d.p.). Using these values, $b = 0.5$ and $g(x)$, we can form the equation:

$$y = 0.081 \cosh\left(\frac{x - 0.5}{0.081}\right) - 0.481 \quad \{0.3 \leq x \leq 0.7\}$$

For the small curves, I will use values of a_s, c_s and $b = 0.85$ for the second (middle) curve and $b = 1.55$ for the third (right-most) curve. Similarly, for the second large curve, I will use values a_l, c_l and $b = 1.2$. This can be shown in the piecewise function $g(x)$ where all values are exact or to 3 decimal places for ease of display:

$$g(x) = \begin{cases} 0.071 \cosh\left(\dfrac{x - 0.15}{0.071}\right) - 0.296, & 0 \leq x \leq 0.3 \\ 0.081 \cosh\left(\dfrac{x - 0.5}{0.081}\right) - 0.481, & 0.3 \leq x \leq 0.7 \\ 0.071 \cosh\left(\dfrac{x - 0.85}{0.071}\right) - 0.296, & 0.7 \leq x \leq 1 \\ 0.081 \cosh\left(\dfrac{x - 1.2}{0.081}\right) - 0.481, & 1 \leq x \leq 1.4 \\ 0.071 \cosh\left(\dfrac{x - 1.55}{0.071}\right) - 0.296, & 1.4 \leq x \leq 1.7 \end{cases}$$

This will give us the graph:

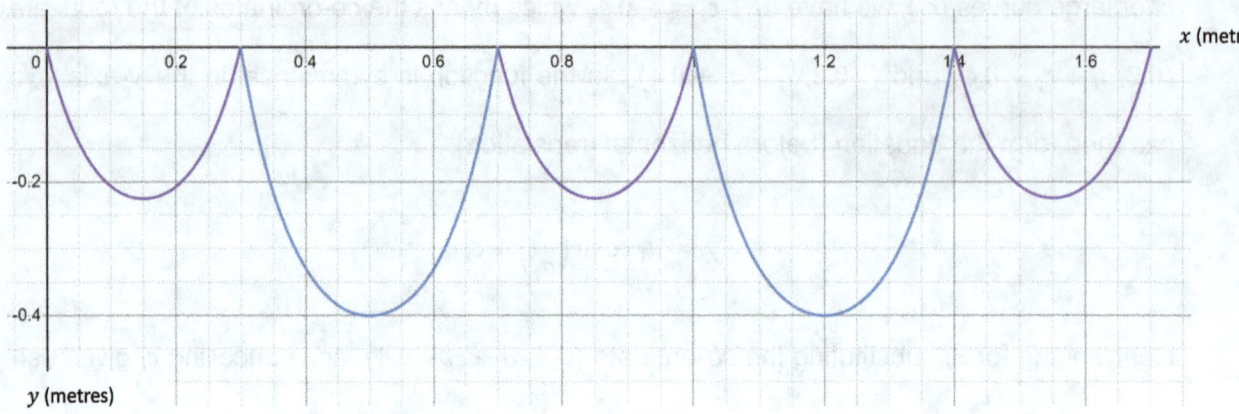

Figure 9: Diwali lights (modelled using catenaries / hyperbolic cosine)

This overall shape of the curves looks slightly more accurate than quadratics when observing hanging chains in real life. Although when both are present in Figure 10 below, the differences between both approaches in terms of rate of change and behaviour near the minimum points (as discussed in Figure 5) can be observed in relation to my designed model.

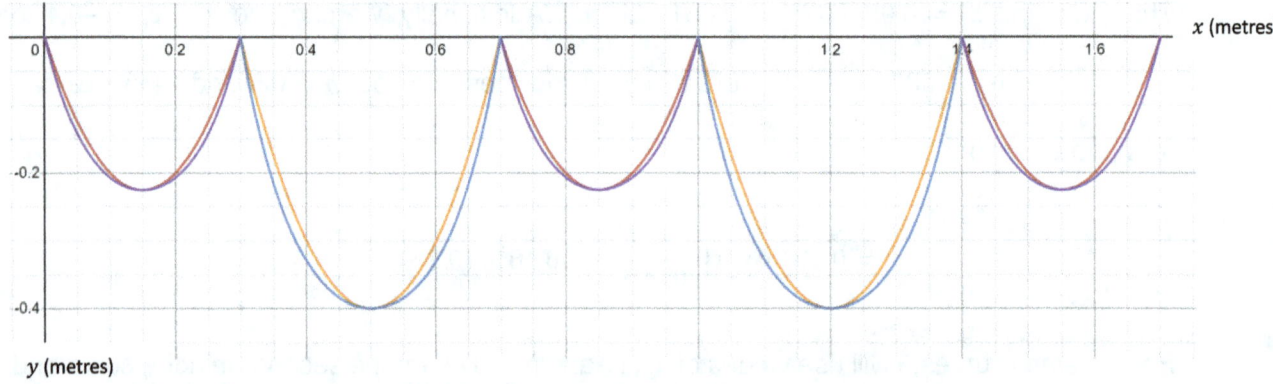

Figure 10: Diwali lights model on wall with both quadratics (red and orange) and hyperbolic cosine (purple and blue)

Both patterns are similar, but using catenary curves, I realised that I would need a little more string light to model the same pattern. Also, the area between the graph and the x-axis is higher. This is due to the slower rate of change around the minimum for the hyperbolic cosine function, making the pattern longer and the area higher. Therefore, this could potentially change the length of string light bought since they come in standard sizes for length, however the difference between both lengths will likely be small. To calculate the lengths and areas I need to achieve the aim, I will derive specific hyperbolic relationships and identities, and also the arc length formula.

Deriving hyperbolic relationships and identities

I will derive the following hyperbolic trigonometric relationships using their exponential forms. The relationship between hyperbolic cosine and hyperbolic sine ($y = \sinh x$) can be shown:

$$\frac{d}{dx}(\cosh x) = \frac{d}{dx}\left(\frac{1}{2}(e^x + e^{-x})\right) = \frac{1}{2}(e^x - e^{-x}) = \sinh x$$

$$\frac{d}{dx}(\sinh x) = \frac{d}{dx}\left(\frac{1}{2}(e^x - e^{-x})\right) = \frac{1}{2}(e^x + e^{-x}) = \cosh x$$

I was surprised when $\frac{d}{dx}(\cosh x)$ was not $-\sinh x$ since $\frac{d}{dx}(\cos x) = -\sin x$ due to similarities in properties. The derivatives remain positive, since hyperbolic functions are formed using exponentials instead of trigonometric functions, as shown above. These will be used with the arc length formula to find the lengths and areas required.

Also, a Pythagorean identity can be derived:

$$\cosh^2 x - \sinh^2 x = \left(\frac{1}{2}(e^x + e^{-x})\right)^2 - \left(\frac{1}{2}(e^x - e^{-x})\right)^2$$

$$= \frac{1}{4}(e^{2x} + e^{-2x} + 2) - \frac{1}{4}(e^{2x} + e^{-2x} - 2)$$

After simplifying, we see:

$$= \frac{1}{2} + \frac{1}{2} = 1$$

This can also be expressed as:

$$\cosh x = \sqrt{1 + \sinh^2 x}$$

(Branson, 2012)

I realised that this hyperbolic identity is not the exact same as the common trigonometric identity $\cos x = \sqrt{1 - \sin^2 x}$ due to the negative sign, as shown by the derivation with exponentials above.

Deriving the length of a curve

I also learnt how to derive and use the arc length formula in this exploration, which is necessary in achieving my aim of finding the length of string light needed for the Diwali lights. Deriving the formula can also give interesting insights on how it is used.

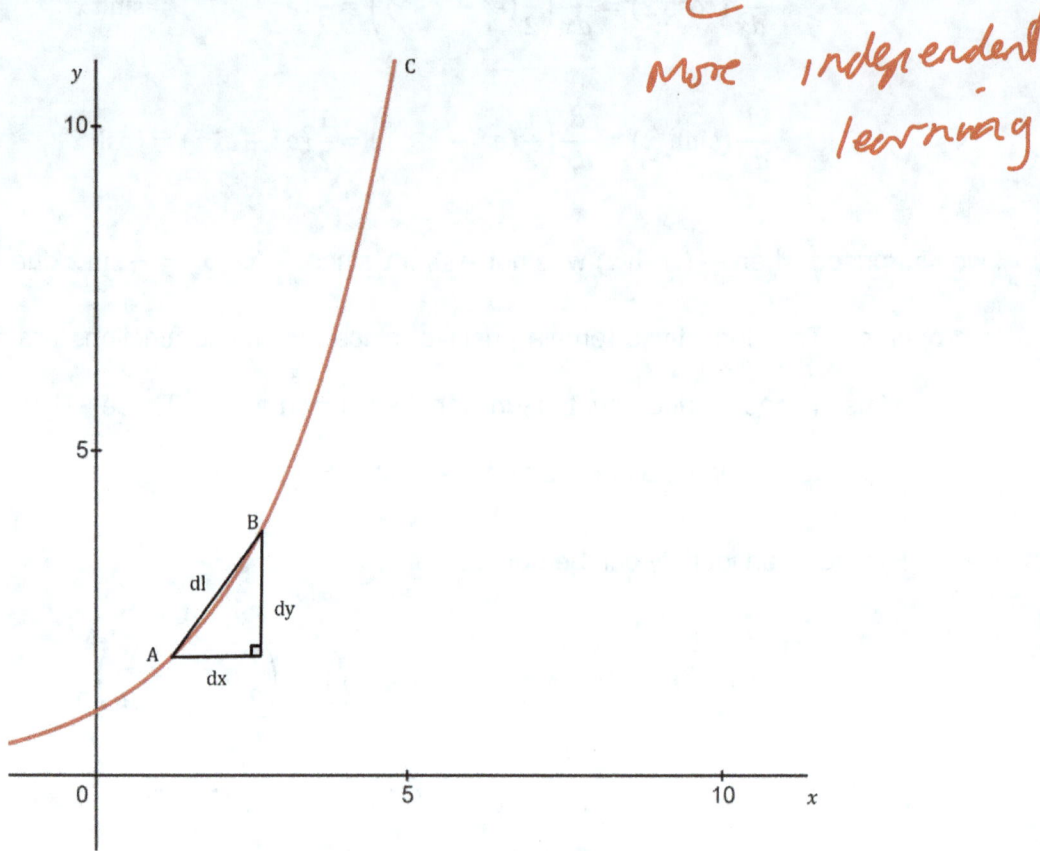

Figure 11: Arc Length Derivation Diagram

Figure 11 shows point A and B on a curve I have drawn, with co-ordinates in the form (x, y). The length between A and B is assumed to be extremely small, although will be presented longer for the purpose of this derivation. Let the distance from A to point C be the arc length, l, which we want to calculate. The small length between A and B can approximate to dl. From A to B, the change in the x-values can be noted as dx and change in the y-values are dy.

(University of Notre Dame, n.d.)

As this is a right-angled triangle, I used Pythagoras' Theorem to derive the arc length. Assuming A to B is an extremely small length:

$$(dl)^2 = (dx)^2 + (dy)^2$$

(University of Notre Dame, n.d.)

Rearranging this equation, we can obtain:

$$dl = \sqrt{(dx)^2 + (dy)^2}$$

$$= \sqrt{(dx)^2 \left(1 + \frac{(dy)^2}{(dx)^2}\right)}$$

$$= \sqrt{\left(1 + \left(\frac{dy}{dx}\right)^2\right)(dx)^2}$$

$$= \sqrt{1 + \left(\frac{dy}{dx}\right)^2}\, dx$$

The arc length can then be shown by the following integral:

$$l = \int_{x_1}^{x_2} dl$$

$$l = \int_{x_1}^{x_2} \sqrt{1 + \left(\frac{dy}{dx}\right)^2}\, dx$$

As $dx \to 0$, I found that the arc length will become more accurate, since A and B are closer, giving a better approximation for total arc length, l. This formula will be used with the polynomials and catenary curves.

useful and well explained derivation of arc length formula

Finding the length of string light needed

Using these derivations, we can find the length of string light needed. To find the length of the first small curve with horizontal width 0.3m, we will apply the arc length formula derived. All calculations are done with the exact values in equations $f(x)$ and $g(x)$, but displayed with 3 decimal places for ease of reading.

We will need to find the first derivative:

$$y = 0.071 \cosh\left(\frac{x - 0.15}{0.071}\right) - 0.296$$

Using derivation $\frac{d}{dx}(\cosh x) = \sinh x$:

$$\frac{dy}{dx} = \sinh\left(\frac{x - 0.15}{0.071}\right)$$

Substituting this into the arc length formula, we can find the arc length of a small curve (l_s):

$$l_s = \int_0^{0.3} \sqrt{1 + \left(\sinh\left(\frac{x - 0.15}{0.071}\right)\right)^2} \, dx$$

We can now use Pythagorean identity derivation to simplify the integral:

$$l_s = \int_0^{0.3} \cosh\left(\frac{x - 0.15}{0.071}\right) dx$$

$$= \left[0.071 \sinh\left(\frac{x - 0.15}{0.071}\right)\right]_0^{0.3}$$

$$= \left(0.071 \sinh\left(\frac{0.3 - 0.15}{0.071}\right)\right) - \left(0.071 \sinh\left(\frac{0 - 0.15}{0.071}\right)\right)$$

$$l_s = 0.575269\ldots \approx 0.575 \text{m} \ (3.d.p).$$

This is the arc length of one small curve. The answer is again rounded to 3 decimal places, since it is an appropriate accuracy to record results to when using a meter stick, as I am rounding to the nearest millimetre. I will round all of the final values to 3 decimal places to maintain consistency with results. By integrating analytically, I also found that performing calculus with hyperbolic functions was similar to calculus with trigonometric functions I have learnt in school.

Following the same process to find the arc length of a large curve (l_l):

$$y = 0.081 \cosh\left(\frac{x - 0.5}{0.081}\right) - 0.481$$

$$\frac{dy}{dx} = \sinh\left(\frac{x - 0.5}{0.081}\right)$$

$$l_l = \int_{0.3}^{0.7} \sqrt{1 + \left(\sinh\left(\frac{x - 0.5}{0.081}\right)\right)^2} \, dx = 0.948401\ldots \approx 0.948\text{m} \ (3\ d.p.)$$

This means the total length of string light needed using the catenary curve model (L_c) can be formed by equation $L_c = 3l_s + 2l_l$.

$$L_c = 3(0.575269\ldots) + 2(0.948401\ldots) = 3.622609 \approx 3.623\text{m} \ (3\ d.p.)$$

The same process has been followed for the quadratic curves in Appendix B, which I found was difficult to fully integrate analytically, so I used a formula which helped me solve an equation in the form $y = \int \sqrt{x^2 + a^2} \, dx$, as it involved integrating a square root without a known identity (eMathZone, n.d.). I found the length using the quadratics, $L_q = 3.554\text{m} \ (3\ d.p.)$. We can see that the length between L_q and L_c is different, where L_c is approximately 7cm, or 0.07m more than L_q (as $L_c - L_q \approx 0.07\text{m}$). Additionally, when viewing the models graphically in Figure 10, the values obtained for length seem appropriate.

We can see that both values calculated for string light length, L_q and L_c, are similar and will not affect the total string light bought (since they come in standard sizes), where I will still buy 4m lights from the shop. We also have to account for the distance between the end of the light pattern to the socket on the wall, a limitation of using both of these models to find the length of string light that needs to be bought. However, I am interested to see the difference between the area underneath the string lights and if the number of flowers I need to buy changes when using both models.

Finding the area underneath the string lights

We can find the area under the string lights to see how many flowers I can place underneath, as shown in Figure 1, where each flower is 0.0025m². Finding the area using catenary curve functions, we can integrate the first small curve:

$$A_s = \int_a^b |y|\, dx$$

$$= \int_0^{0.3} \left| 0.071 \cosh\left(\frac{x - 0.15}{0.071}\right) - 0.296 \right| dx$$

$$= \left[\left| 0.071^2 \sinh\left(\frac{x - 0.15}{0.071}\right) - 0.296x \right|\right]_0^{0.3}$$

$$= \left|\left(0.071^2 \sinh\left(\frac{0.3 - 0.15}{0.071}\right) - 0.296(0.3)\right) - \left(0.071^2 \sinh\left(\frac{0 - 0.15}{0.071}\right)\right)\right|$$

$$A_s = 0.047858 \ldots \approx 0.048 \text{m}^2 \; (3\, d.p.)$$

[Handwritten annotation: More clear working toward aim]

This is the area between one small curve and the x-axis. Using the same process for larger curves:

$$A_l = \int_{0.3}^{0.7} \left| 0.081 \cosh\left(\frac{x - 0.5}{0.081}\right) - 0.481 \right| dx = 0.115531 \ldots \approx 0.116 \text{m}^2 \; (3\, d.p.)$$

This means the total area between the catenary curves and the x-axis can be calculated: $3(A_s) + 2(A_l) = 0.374640 \ldots \approx 0.375 \text{m}^2 \; (3\, d.p.)$. I then found the area using quadratics, shown in Appendix C, as it is a repetitive process. The total area between the quadratics and the x-axis was $0.348 \text{m}^2 \; (3\, d.p.)$.

The area I have space for decorations in my house is 1.7m wide and 1.0m tall on a section of my living room wall. I want to place the string lights of maximum length 3.623m (using catenary curves) at the top of this area, meaning the top will be the x-axis. The area of the available space for all decorations is $1.7 \times 1.0 = 1.7\text{m}^2$. Finding the area underneath the string lights, we get $1.7 - 0.348333\ldots = 1.351666\ldots \approx 1.352\text{m}^2$ ($3\ d.p.$) when using the quadratic curves. Each flower has an area of 0.0025m^2, so dividing this value by the area calculated gives us $\frac{1.351666\ldots}{0.0025} = 540.6664\ldots$, meaning 540 of these identical flowers will fit on the area I want to decorate, when using the quadratic model.

Furthermore, $1.7 - 0.374640\ldots = 1.325359\ldots \approx 1.325\text{m}^2$ ($3\ d.p.$) gives us the area under the lights for the hyperbolic cosine model. As $\frac{1.325359\ldots}{0.0025} = 530.143817\ldots$, 530 flowers will fit on the area. These values make sense as less flowers will be needed when the catenary curve model is used, as the area underneath the string lights was slightly less, which we can also see using Figure 10. This means I will need to buy a pack of 500 flowers regardless of the model used when they are arranged in a grid pattern, which was surprisingly more than I expected. Using the quadratic model, I would need to buy a pack of 40 more flowers, but when using the hyperbolic cosine model, I will need to only buy 30 more.

A limitation is that these calculated values are likely to be an overestimate for the number of flowers which can be placed on the wall. This is because the flowers are arranged in a grid and the area calculated (under the lights) does not account for intersections between the curves and the grid, meaning I may have to remove flowers which are overlapped by the lights. However, this overestimate is minor, due to there being a large number of flowers placed on the wall, meaning the packs of flowers I have to buy will likely not change due to this.

D+ critical reflection

Summary Table

All values calculated relevant to the aim can be shown below.

	Using Quadratic function	Using Hyperbolic Cosine function
Total Length of string light needed (m)	3.554	3.623
Area covered by flowers (m²)	1.352	1.325
Number of flowers needed	540	530

Table 2: Values for length and area found from exploration, rounded to 3 d.p. when appropriate.

Conclusion

I have fulfilled my aim of calculating the length of string light needed to form the designed pattern and I have also found how many flowers I can place by calculating the area under the lights. I achieved this using quadratics, but then realised catenary curves formed by hyperbolic cosine functions were a better representation of this pattern in real life.

Integrating analytically was interesting when finding the length of string lights needed since I had never seen hyperbolic functions before, and I had to derive a Pythagorean Identity to simplify these. Using techniques like the arc length formula with these functions was challenging at times, yet it was fascinating to see the similarities and differences between hyperbolic and standard trigonometric functions. I also used my GDC and Desmos to effectively display and perform some of the graphs and calculations.

Others can also use the information found in this exploration to plan out and design their own decorations, either for Diwali or other festivals or celebrations. It can provide them with the exact length of festive string light they would need if they used the specifications, along with the area left for other decorations on their wall, such as flowers.

A difficulty with my exploration was finding the arc length using quadratics, as I struggled to integrate it analytically the whole time. I found a formula online to solve equations in the form $y = \int \sqrt{x^2 + a^2}\, dx$, which I used to obtain the final result, seen in Appendix B.

An extension to this exploration is modelling these decorations using a different approach, specifically using standard trigonometric functions, so I can further compare the differences between specific trigonometric and hyperbolic functions. If I used identical curves in the specification to model the pattern with the lights, I could repeat the pattern infinitely using trigonometric functions such as $y = \sin x$ or $y = \cos x$. This would be a better method than using parabolas since they are continuous functions on their natural domains.

good conclusion

more reflection

possible extensions and applications mentioned

Reference List

Branson, J. (2012, October 21). *Hyperbolic Function Identities*. Retrieved November 25, 2023, from University of California San Diego: https://hepweb.ucsd.edu/ph110b/110b_notes/node49.html

Carlson, S. C. (2017, Febuary 3). *catenary*. Retrieved October 18, 2023, from Encyclopedia Britannica: https://www.britannica.com/science/catenary

Desmos. (n.d.). *Desmos Graphing Calculator*. Retrieved November 20, 2023, from Desmos: https://www.desmos.com/calculator

eMathZone. (n.d.). *Integration of the Square Root of a^2+x^2*. Retrieved October 20, 2023, from eMathZone: https://www.emathzone.com/tutorials/calculus/integration-of-square-root-of-a2x2.html

Georgia Institute of Technology. (2018, April 15). *The shape of a hanging chain*. Retrieved November 25, 2023, from Georgia Institute of Technology: https://mccuan.math.gatech.edu/courses/7581/chainreport.pdf

Newcastle University. (n.d.). *Euler's Formula and Euler's Identity*. Retrieved November 25, 2023, from Newcastle University: https://www.ncl.ac.uk/webtemplate/ask-assets/external/maths-resources/core-mathematics/pure-maths/algebra/euler-s-formula-and-euler-s-identity.html

Rawpixel. (n.d.). *Marigold png, flower sticker, transparent background*. Retrieved December 1, 2023, from Rawpixel: https://www.rawpixel.com/image/6201859/png-flower-sticker

University of Notre Dame. (n.d.). *Lecture 16: Arc Length*. Retrieved November 19, 2023, from University of Notre Dame: https://www3.nd.edu/~apilking/Calculus2Resources/Lecture%2016/NotesL16.pdf

Whitman College. (n.d.). *4.11 Hyperbolic Functions*. Retrieved November 25, 2023, from Whitman College: https://www.whitman.edu/mathematics/calculus_online/section04.11.html

Appendices

Appendix A

We can use Euler's formula $e^{ix} = \cos x + i \sin x$ to derive $\cos x = \frac{e^{ix}+e^{-ix}}{2}$ (Newcastle University, n.d.). We can see:

$$e^{-ix} = \cos(-x) + i\sin(-x)$$

As $\cos x$ is an even function and $\sin x$ is odd:

$$e^{-ix} = \cos x - i\sin x$$

Adding e^{ix} to e^{-ix}, we get:

$$e^{ix} + e^{-ix} = (\cos x + i\sin x) + (\cos x - i\sin x)$$

$$e^{ix} + e^{-ix} = 2\cos x$$

We can then see:

$$\cos x = \frac{e^{ix} + e^{-ix}}{2}$$

(Newcastle University, n.d.)

Appendix B

To find the arc length using the quadratic curves, we can use the first small curve:

$$y = 10(x)(x - 0.3) = 10x^2 - 3x$$

$$\frac{dy}{dx} = 20x - 3$$

Using the arc length formula:

$$l_s = \int_0^{0.3} \sqrt{1 + (20x - 3)^2}\, dx$$

$$= \int_0^{0.3} \sqrt{400x^2 - 120x + 10}\, dx$$

$$= \int_0^{0.3} \sqrt{400} \sqrt{x^2 - \frac{3}{10}x + \frac{1}{40}}\, dx$$

Through removing the constant from the integral and completing the square:

$$= 20 \int_0^{0.3} \sqrt{\left(x - \frac{3}{20}\right)^2 + \frac{1}{400}}\, dx$$

Using the formula: $\int \sqrt{x^2 + a^2}\, dx = \frac{x}{2}\sqrt{x^2 + a^2} + a^2 \ln\left|x + \sqrt{x^2 + a^2}\right| + C$ (eMathZone, n.d.):

$$= 20 \left[\frac{(x - 0.15)}{2} \sqrt{(x - 0.15)^2 + 0.0025^2} + 0.0025^2 \ln\left|(x - 0.15) + \sqrt{(x - 0.15)^2 + 0.0025^2}\right| \right]_0^{0.3}$$

$$l_s = 0.565263 \ldots \approx 0.565\text{m} \ (3\ d.p.)$$

Following the same process for the first large curve and using the arc length formula:

$$y = 10(x - 0.3)(x - 0.7) = 10x^2 - 10x + 2.1$$

$$\frac{dy}{dx} = 20x - 10$$

$$l_l = \int_{0.3}^{0.7} \sqrt{1 + (20x - 10)^2}\, dx = 20 \int_{0.3}^{0.7} \sqrt{\left(x - \frac{1}{2}\right)^2 + \frac{1}{400}}\, dx$$

$$l_l = 0.929356 \ldots \approx 0.929\text{m} \ (3\ d.p.)$$

This means the total length of string light needed using quadratics, L_q, is:

$$L_q = 3(0.56526\ldots) + 2(0.929356\ldots) = 3.554492 \approx 3.554\text{m} \ (3\ d.p.).$$

Appendix C

Using the first small quadratic curves after expanding the brackets of the function:

$$A_s = \int_0^{0.3} |10x^2 - 3x|\, dx = \left[\left[\frac{10}{3}x^3 - \frac{3}{2}x^2\right]\right]_0^{0.3} = \left|\left(\frac{10}{3}(0.3)^3 - \frac{3}{2}(0.3)^2\right) - (0)\right| = 0.045\text{m}^2$$

Similarly for the larger curves:

$$A_l = \int_{0.3}^{0.7} |10x^2 - 10x + 2.1|\, dx = 0.106666\ldots \approx 0.107\text{m}^2 \ (3\ d.p.)$$

This means the total area between the curves and the x-axis is:

$3(A_s) + 2(A_l) = 0.348333\ldots \approx 0.348\text{m}^2\ (3\ d.p.)$.